Cybersecurity and Traceability in Factory 4.0

SCIENCES

Systems and Industrial Engineering,
Field Director – Jean-Paul Bourrières

Industrial and Productive Engineering,
Subject Head – Samir Lamouri

Cybersecurity and Traceability in Factory 4.0

Blockchain Assets

Patrick Sondi
Valentin Mullet

ISTE

WILEY

First published 2026 in Great Britain and the United States by ISTE Ltd and John Wiley & Sons, Inc.

ISTE Ltd
27-37 St George's Road
London SW19 4EU
UK

www.iste.co.uk

John Wiley & Sons, Inc.
111 River Street
Hoboken, NJ 07030
USA

www.wiley.com

Library of Congress Control Number: 2026938344

British Library Cataloguing-in-Publication Data
A CIP record for this book is available from the British Library
ISBN 978-1-78945-224-2

ERC code:
PE6 Computer Science and Informatics
PE6_5 Security, privacy, cryptology, quantum cryptography

The manufacturer's authorized representative according to the EU General Product Safety Regulation is Wiley-VCH GmbH, Boschstr. 12, 69469 Weinheim, Germany, e-mail: Product_Safety@wiley.com.

Contents

Preface

Intention, abstract and target audience for this work

This book proposes another viewpoint on the research work carried out in collaboration by the authors for a doctoral thesis, namely, with Valentin Mullet, author of the thesis and doctor in computer science from the University of the Littoral Côte d'Opale, and Patrick Sondi, thesis director and professor at the IMT Nord Europe's Center for Education, Research and Innovation in Digital Systems (CERI SN-Institut Mines-Telecom Nord Europe).

The authors' intention is to better understand the crucial challenges involved with the digitalization of production plants in the context of Industry 4.0. Besides the need to recall that the productivity gains brought about by the introduction of new digital technologies in factories involve a number of risks, constraints, dependencies and new costs inherent to the factories themselves, the authors' main goal is to share the idea that there are practical ways manufacturers can approach this transition in a way that remains truly beneficial for their overall activity.

The Internet of Things and cloud computing enable online interactions between third parties, such as customers and suppliers, and the traceability system of a factory. The industrial blockchain implements a paradigm that makes it possible to set up a tamper-proof ledger of transactions between such partners, who are individually free to record actions in it and verify the actions of others in an ad hoc way. In this regard, reliable, transparent and directly verifiable traceability can be put in place by each partner (supplier/factory, factory/subcontractor, factory/customer, etc.). In the era of connected factories, it therefore represents a solution that can be implemented between the factory information system and the various partners in order to automate traceability management. Nevertheless, the overall traceability system of the plant may include internal components with strong implications on confidentiality and whose link with traceability is merely the responsibility of the factory and does not concern its partners. Within this context, the resulting transparency would be gained at the expense of data confidentiality. The purpose of the work described in this book is to develop an architecture and to propose evolutions of traceability systems to allow the integration of a blockchain-based solution intended to ensure traceability between the factory and the partners without compromising confidentiality on the grounds of transparency.

This book, accessible to any public concerned with Industry 4.0, is in particular intended for students, researchers and operational engineers in the fields of IT, cybersecurity and industrial engineering who engage in training or projects for the digitalization of industry according to the Industry 4.0 paradigm. More specifically, it provides a different point of view on the scientific literature related to these fields, concrete solutions and tools for the development of practices in terms of design, implementation and evaluation of blockchain-based traceability systems that prove to be secure, efficient in terms of data volume and energy consumption, without increasing the risks to the sustainability of industrial activity.

Book organization

The book was written entirely by the two authors, based on authors' notes on traceability projects prior to the work of the thesis, the thesis manuscript and translations of scientific articles, all reorganized and rewritten in order to ensure the coherence of the proposition made in the book. It comprises five chapters.

Chapter 1 addresses the challenges related to the cybersecurity of the technologies introduced by Industry 4.0 and the relationships between these technologies and the traceability system of an Industry 4.0 factory. This global vision of the factory from a cybersecurity point of view makes it possible to better understand the concepts presented in detail in the following chapters of the book. A synthesis of innovative solutions and best practices relating to cybersecurity at the global level for an Industry 4.0 factory is proposed in the second part of this chapter with a view to strengthening awareness of these issues in digitalization projects within an industrial environment.

In order to refine this broad concept that traceability constitutes, Chapter 2 provides a general overview of the proposed models before focusing on the notion of product-oriented traceability that prevails in manufacturing production plants, which have been the focus of our studies. An overview of the standards and contributions of the scientific and technical literature on traceability in a factory 4.0 is proposed, as well as our own approach to the characterization of traceability data, its collection and its processing in adherence to clearly defined and motivated objectives.

After this introduction, the reader will be ready for the discovery of blockchain technology in Chapter 3, and understand how it can be integrated into the implementation of traceability in a factory that aims to follow the guidelines of Industry 4.0. This presentation also introduces a description of the architecture that we propose for this purpose, in particular aiming at preserving the confidentiality of sensitive data without jeopardizing the inherent transparency of blockchain technology, as well as its implementation using an *open source* blockchain platform, in this case MultiChain. The twofold merit of this choice resides in the fact that our contribution is fully accessible to be used as the working foundation for real blockchain implementation projects, and that it enables the reproducibility of all the experiments and results presented in the book.

Since blockchain is a recent technology and surrounded by many fantasies about its efficiency and its presumed impacts, especially in terms of energy consumption, we wanted to equip project leaders with a tool that enables them to perform a preliminary evaluation through simulation before any investment. Therefore, Chapter 4 presents a series of models based on discrete event simulation for the creation of a simulator reproducing the operation of a factory that integrates a blockchain-based traceability system. A case study, built through the implementation of our architecture and our contributions, is proposed to support the description of the operation and use of the simulator. The evaluation of different production scenarios from a number of plant configurations makes it possible to underline the impact of the blockchain mechanisms proposed for traceability on the operation of the plant based on criteria of interest such as the volume of traceability data, the impact of their storage, the impact on the energy consumption or the environmental impact of the plant resulting from the use of these solutions. For us this is the opportunity to analyze these results to propose new models for the distribution of the impacts related to the introduction of blockchain between the factory, its suppliers and its customers, bearing in mind that this technology provides a service that is as profitable to the factory as to these other actors.

Finally, Chapter 5 is an overview of the work and results presented, as well as of the open avenues that we conceive around the digitalization of factories in the context of Industry 4.0 and beyond.

May 2026

List of Acronyms

APT Advanced Persistent Threats

B2B Business to Business

B2C Business to Customer

BPCAT Blockchain based Privacy-preserving and product-Centered Approach for Traceability

C2C Customer to Customer

CFDA China Food and Drug Administration

CPSs Cyber-Physical Systems

DCSs Distributed Control Systems

DDoS Distributed Denial-of-Service

DEVS Discrete Event System Specification

DMZ Demilitarized Zone

EAN European Article Numbering

EIF Ensemble Intelligence Framework

ERP Enterprise Resource Planning

GDPR General Data Protection Regulation

HACCP Hazard Analysis Critical Control Points

HMI Human–Machine Interface

ICSs Industrial Control Systems

IDSs Intrusion Detection Systems

IEC International Electrotechnical Commission

IED Intelligent Electronic Device

IoT Internet of Things

IS Information System

ISO International Organization for Standardization

IT Information Technology

LAN Local Area Network

LPWAN Low Power Wide Area Network

MDSEA Model-Driven Service Engineering Architecture

MES Manufacturing Execution System

MSED Manufacturing Security Enforcement Device

NNO Neural Network Oracle

OCR Optical Character Recognition

OPC UA Open Platform Communication Unified Architecture

OS Operating System

OT Operational Technology

P2P Peer-to-Peer

PBFT Practical Byzantine Fault Tolerance

PLC Programmable Logic Controller

PoET Proof of Elapsed Time

POS Proof of Stake

POW Proof of Work

RTU Remote Terminal Unit

SCADA Supervisory Control And Data Acquisition

SDN Software-Defined Networking

SMO Smart Manufacturing Objects

TEE Trusted Execution Environment

TRU Traceable Resource Units

UPC Universal Product Code

WLAN Wireless LAN

List of Figures

Chapter 1

Chapter 2

Chapter 3

Chapter 4

List of Tables

Chapter 1

Chapter 2

Chapter 3

Chapter 4

1

Challenges Involving Cybersecurity and Traceability in Factory 4.0

1.1. Introduction

The industrial sector has witnessed several revolutions over the last few centuries (Figure 1.1). The first occurred with mechanization, introduced by the steam engine, in 1765. Driven by electricity and oil production in 1870, mass production brought about the second revolution through the development of high-tech industries such as automobiles and aeronautics. The third revolution was that of automated production in 1969, supported by electronics with the advent of the transistor and the microprocessor; it subsequently initiated the appearance of programmable logic controllers (PLCs), robots and computer technologies in factories.

In 2011, a fourth phase was conceptualized in Germany as an evolution of traditional factories (Kagermann et al. 2013) toward more flexibility and adaptation to constantly developing production environments: the Industry 4.0 paradigm, also known as the Industrial Internet of Things (IoT) or Industrial Internet, was born. For example, in the agri-food sector, the goal of the Industry 4.0 paradigm is to connect farms and product processing plants to the Internet in order to improve overall efficiency and productivity (approximately 15%–20% targeted).

On a more general level, the necessary hyperconnectivity for this operation must allow the collection of large volumes of data originating from the value chain to be used for a wide range of purposes, such as:

For a color version of all the figures in this chapter, see: www.iste.co.uk/sondimullet/cybersecurity.zip

– exchange of information in real time between devices belonging to factories, suppliers or customers;

– data acquisition and storage for traceability and computing performance management;

– data processing for predictive maintenance or remote monitoring to reduce machine downtime;

– task automation and stock reduction;

– improved service levels and product quality.

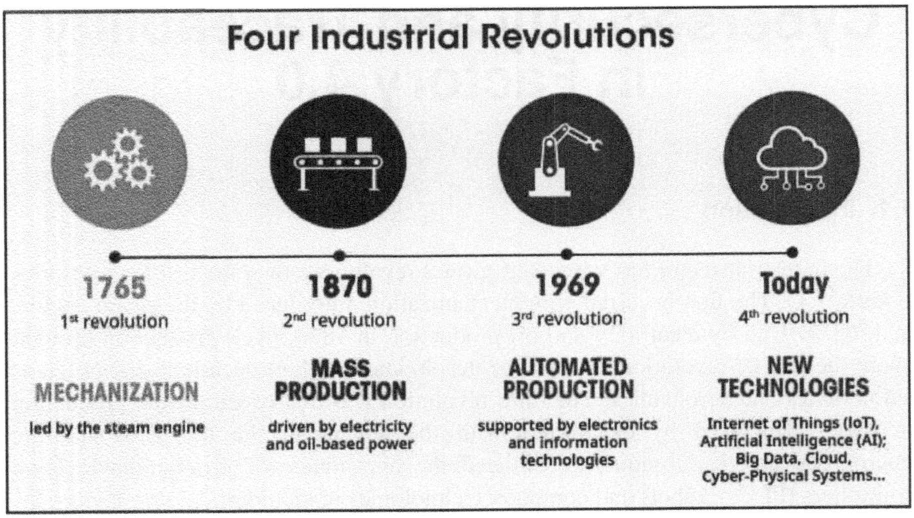

Figure 1.1. *Timeline of the industrial revolutions (source: Geandarme 2018)*

The implementation of Industry 4.0 consists on the one hand of collecting as much information as possible about industrial activity in order to digitize it, process it and aggregate it with other information. This digital transformation or digitalization is materialized by the introduction in the factory of innovative technological objects such as the IoT, especially for collecting data and industrial control, Cloud computing for the processing and storage of these massive amounts of data (Big Data) and optimization tools using artificial intelligence (AI) for the automation of factory management. However, it brings about several drawbacks that are imported in parallel into the industrial activity, namely problems inherent in new technologies in terms of operational safety and cybersecurity, as well as the creation of new dependencies likely to undermine the resilience of industrial activity itself. Nevertheless, these technological objects are necessary to connect operators and machines or equipment with production management tools in order to adapt to customer requests and provide a quick response to the slightest variation. Factory 4.0 can thus achieve a high level of control customization in a context that mainly remains that of mass production. As such, innovative objects are involved in the creation of a significant part of the value and in improving the quality and standards of the factory, de facto justifying their adoption.

On the other hand, working in Factory 4.0 is perceived differently due to the implementation of combined processes between machines and operators. The global challenges of developing Industry 4.0 therefore also include, in addition to the technical means themselves, equipment and training specifically oriented for operators in connection with new technologies. As an example, we can mention all the developments in augmented reality (AR) and virtual reality (VR) that allow new forms of interaction between the operator, the production system and the control system. These aspects are an integral part of a fifth stage of the evolution of the industry initiated by the European Union (European Commission 2021) referred to as Industry 5.0 but will not be specifically addressed in this book. They highlight however that the development of the industry of the future requires a balanced distribution of resources between operator training, investment in equipment and data protection.

1.2. Traceability challenges in Factory 4.0

Consequently, although the central subject of this book is traceability in Industry 4.0, the hyperconnectivity that characterizes this context implies that this traceability be outlined and addressed in conjunction with their relationship to cybersecurity.

Considered for a long time a response to regulatory and commercial constraints or even a marginal resource in technical diagnostics, traceability has become a driver for performance improvement within the context of the digital transformation of factories. The new technologies of Industry 4.0 are actually based on the use of collected and historical data, of which a significant part is created by industrial traceability. With the increasing digitization of industry, traceability is now part of all production units within a factory. All stages, from design to delivery, including production processes, are involved in data collection, processing and analysis.

1.2.1. *The volume of data*

Nonetheless, this generalization causes the volumes of data generated by traceability processes to often exceed the storage capacities of a factory. In a report published in 2018 (Reinsel et al. 2018), IDC (International Data Corporation), in partnership with Seagate, proposed in its white paper a study comparing the amount of data (called "the datasphere") generated by different professional sectors (Figure 1.2). The share of the manufacturing industry is the largest with 3584 Eo (10^{18} bytes) due to its maturity, the industry's investment in the IoT and its ability to produce 24 hours a day and 7 days a week.

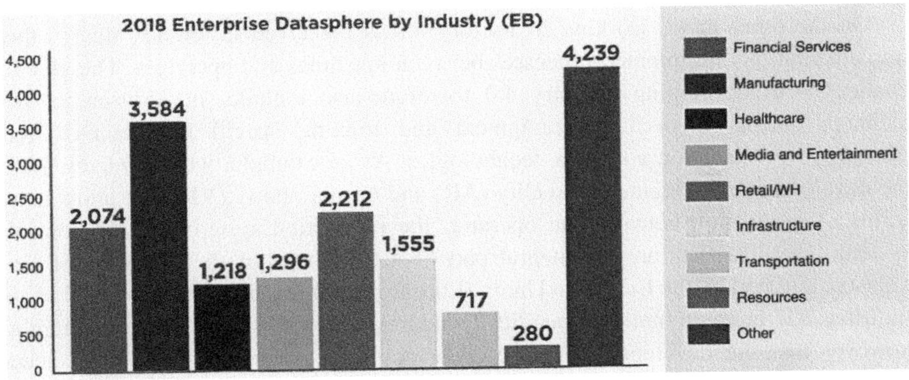

Figure 1.2. *Datasphere of companies classified by industry (source: Reinsel et al. 2018)*

To address this problem, plant managers are resorting to outsourcing the storage of their data to the Cloud. This presents several major risks such as Internet dependency, exposure of the factory IT system to the Internet, loss of control over data access management and even geopolitical risks. It is therefore necessary that the management of this problem of data volume be addressed in a more global context including cybersecurity considerations.

1.2.2. *Transparency*

For companies, the supply chain is a fundamental area that consists of tracking the origin and history of products from beginning to end. However, its opacity and lack of data recording were seen as major issues. This opacity was due in part to the obsolescence of the system, which was still too often based on paper-based registers requiring manual handling that therefore was prone to errors, loss of time and loss of information because of its volatility. The digitalization of traceability introduced by Industry 4.0 provides, in part, a solution to this problem.

Product quality and origin are indicators whose importance is emphasized by companies as guarantees of brand image and meeting customer needs. Given that the need for visibility is thus more pressing, companies are compelled to improve the transparency and traceability of their supply chain. In this context, the notion of transparency focuses on mapping the entire supply chain and traceability examines individual batches of components or purchase orders as they progress through the supply chain. Another question of vigilance for these companies is the fight against counterfeiting, which finds in this system of total traceability an effective tool to guarantee product authenticity. In this regard, traceability represents an investigation tool for identifying what causes defects in products or inefficiencies in production processes. When it includes several actors of the supply chain, in particular suppliers, customers, regulators and other potential auditors, data integrity and open access must be guaranteed to each actor under well-defined conditions.

The organization of the traceability system for ensuring this transparency, in addition to integrity and open access to data, has many implications on factory cybersecurity. In addition, the need for transparency should not be satisfied at the expense of another aspect, which is confidentiality. Indeed, data generated by production processes may contain critical information relating to the know-how or other crucial interests of the factory and which cannot therefore be exposed at the risk of jeopardizing the company's competitiveness. This is all the more critical as the volume of data is almost systematically the main factor for resorting to the use of Cloud computing for the storage of at least part of it. A balance must therefore be found, in connection with cybersecurity, in order to be able to guarantee transparency, integrity and compliance with confidentiality within the traceability system for the data to which it applies.

1.2.3. *The contribution of blockchain*

Blockchain is one of the components of Industry 4.0 and it provides a set of functionalities that enables the implementation of a historical data register while guaranteeing transparency, integrity and free access to participants. Part of the work presented in this book consists of adding mechanisms that guarantee confidentiality to the data that requires it. In this case, the aim of the research covered throughout this book is to reach the following objectives:

1) develop the conventional factory traceability system toward the blockchain by covering the supply chain, from the supplier up to the customer;

2) guarantee the integrity, transparency and confidentiality of critical data within this system;

3) achieve a successful assessment of the impacts of this evolution from the traceability system to the blockchain by means of a simulation tool that could also be used as a tool for evaluations prior to development planning.

Nevertheless, for a better understanding of the issues related to the implementation of this blockchain-based traceability, as well as its outlines relating to the cybersecurity of a factory 4.0, we will first present a generic description framework for a factory, as well as a description of cybersecurity based on this framework (Mullet et al. 2021). The studies concerning traceability and its implementation using a blockchain will be the subject of the following chapters.

1.3. Plant overview

In order to better understand the notion of cybersecurity within an industrial environment, it should be recalled that the factory comprises several departments with different requirements and work processes. We first propose a subdivision of the plant into six generic zones that we call perimeters.

1.3.1. *Definition of perimeters*

1.3.1.1. *Manufacturing production*

This is the main zone of a factory where the production lines are located, each allocated to some of the many stages required to manufacture final products. The devices existing in this perimeter belong to two groups: Industrial Control Systems (ICSs) and Cyber-Physical Systems (CPSs).

The ICS category mainly encompasses control components that act together to achieve an industry goal (Lezzi et al. 2018). These systems have been in use since the second half of the 20th century (Sicard et al. 2019). Within this category, we can mention the Programmable Logic Controller (PLC), the Remote Terminal Unit (RTU) and the Intelligent Electronic Device (IED) (Rubio et al. 2019). To interact with the hardware controller and access the data collected by the ICS environment, administrators utilize a Human–Machine Interface (HMI), which is also employed to display the device status (Asghar et al. 2019).

The second group, CPSs, is an umbrella for anything that integrates computing operations, networking or physical processes. These systems enable interaction between the digital world and the physical world. As an example, a manufacturing line can be thought of as a CPS (Lezzi et al. 2018). This is because they use sensors and other embedded systems to collect data from physical processes. Several authors (Lu et al. 2014; Lezzi et al. 2018) show that ICSs are a field of application for CPSs.

1.3.1.2. *Logistics zones*

Industrial logistics differs from traditional logistics in that it is adapted to the management of production flows, thus widening its field of action. Its perimeter consists of multiple structures such as workshops and warehouses where forklifts are used. In this field where mobility is a key point, the devices used most are wireless devices (barcode scanners and tablets).

1.3.1.3. *Active supervision*

Active supervision represents tertiary activity, which covers the factory offices. It includes departments such as accounting, sales, human resources (HR) and the local IT system (IT). The most common devices are desktops, smartphones, monitors and printers.

1.3.1.4. *Research and development*

Research and development refers to laboratories and offices involved in innovation and the development of new products or services. This is the first step in the development process, which means that some equipment is new or experimental. Engineers use devices that are more powerful than common desktop computers and usually require specific configuration.

1.3.1.5. *Living zones*

The living zone corresponds to the rooms where employees gather (canteen, meeting rooms, rest rooms, etc.). These zones come with VoIP phones, tablets, screens, etc.

1.3.1.6. *External zones*

The external zone refers to everything located outside the factory. This includes physical places such as parking lots, as well as virtual places such as the Internet (Cloud, etc.).

1.3.2. *Interactions between perimeters*

The perimeters play a critical role in the industrial manufacturing process and their interaction at the IS level is mandatory, as can be seen in Table 1.1.

1.3.3. *Network perimeters in the factory*

Several types of ICSss can be found in the manufacturing sector (Asghar et al. 2019). These ICSs are categorized into two layers: the physical control layer and the logical control layer. This perimeter includes three subnetworks: the Supervisory Control And Data Acquisition (SCADA) system, Distributed Control Systems (DCSs) and programable logic controllers/sensors/protocols. Figure 1.3 depicts these different networks with the equipment and communication protocols used.

	Presence of computer interactions					
	Manufacturing-production	Logistics	Active supervision	Research and development	Living area	External area
Manufacturing-production	X	X	X	X		
Logistics	X	X	X	X	X	
Active supervision	X	X	X	X	X	X
Research and development	X		X	X		
Living areas			X		X	X
External areas			X		X	X

Table 1.1. *IT interactions between plant perimeters*

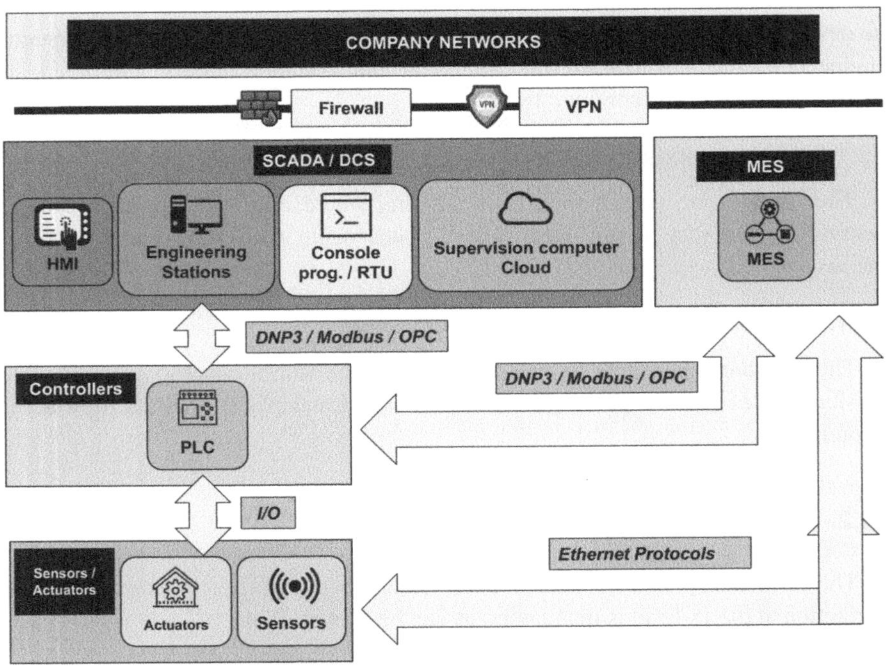

Figure 1.3. *Overview of the manufacturing-production perimeter networks*

1.3.3.1. *SCADA*

A SCADA system enables the supervision of data acquisition and monitoring of the production system. It is also used for the remote control of sites by administrators adopting a centralized control system.

1.3.3.2. *DCS*

A DCS is composed of autonomous controllers installed in a manufacturing or production unit. The DCS system uses these controllers to remotely supervise and monitor a unit. The difference between DCS and SCADA systems is that the SCADA is adapted to individual site control due to its centralized nature, while DCSs can control complex processes that can span multiple sites because of its distributed network characteristics.

1.3.3.3. *PLC/sensors/protocols*

All control devices such as PLCs, sensors or protocols (DNP3/Modbus distributed network protocols) are located in the latter subnet, which is integrated into the physical control layer.

A PLC is the logical interface between SCADA and DCS systems. The PLC is designed to receive control commands and send back the status of the sensors. To establish the connection between the PLC and the SCADA (Abu Jassar et al. 2022), specific communication protocols have been designed by industrial system manufacturers (DNP3/Modbus/OPC).

1.3.4. *Access controls and perimeter monitoring*

From the moment interactions take place between perimeters, precise access controls have to be performed according to the information exchanged, which can be very sensitive. Another important aspect of security is monitoring, which allows users to be notified when a security breach occurs or when the system is under attack. Table 1.2 shows the different types of access controls available in the literature for the defined perimeters, as well as some monitoring solutions applicable to manufacturing plants.

1.4. Overview of cybersecurity in Industry 4.0

Industry 4.0 represents a revolution in the industrial world through the introduction of disruptive technologies such as the IoT and cloud computing within the factory. The resulting increased automation and improved production synergy between stock, supply chains and customer demands are accompanied by threats and attacks originating from the Internet. Despite the extensive literature existing on the topic of cybersecurity, many industrial players are only just beginning to realize the impact of cybersecurity on the preservation of their business. This part of the book will be dedicated to the presentation of the concepts and practical aspects related to the cybersecurity in an Industry 4.0 manufacturing plant. It focuses on the aspects of cybersecurity related to traceability, but additional understanding can be gained by reading Flaus (2019).

In order to create this intelligent production environment, innovative technologies are needed to manage autonomous communications between every industrial device in the factory and the Internet. These technologies (Alcácer and Cruz-Machado 2019) include the IoT, cloud computing, big data, digital twins, AR, 3D printing, AI and new generation CPSs, as illustrated in Figure 1.4.

Plant perimeters	Equipment used	Access methods	Monitoring methods	Types of networks
Manufacturing-production	PLC, RTU, IED, HMI	Username/Password or badge access (human), Device ID via public and private keys (machines)	Video surveillance, traceability verification, logging, remote administration (SNMP or integrated)	SCADA, DCS
Logistics	Mobile terminals (tablets, barcode scanners, laptops)	Badge access (for humans), public/private keys (machines)	Video surveillance, system information control, logging, remote administration (IDS/IPS probes)	Wireless
Active supervision	Office automation, printers, laptops, smartphones	Password, badge access (human), network or system identification (terminals)	System controls, logging, remote administration (SNMP)	Office
Research and development	Workstations, experimental devices	Badge or biometric access (human), network	System verifications (terminals) and system checks, logging, remote administration (SNMP)	LAB
Living areas	VoIP telephony, wireless displays, tablets	No specific control (human), system information control (terminals)	Video surveillance, remote administration (SNMP or *built-in*)	Office
Physical external areas	Cameras	Badge access (human), system and network controls (terminals)	Logging, video surveillance, remote administration (SNMP, etc.)	CAM
Virtual external zones	Cloud service providers	ID/password (human), system and rules, firewalls (devices)	Logging, traffic monitoring, intrusion prevention (IPS) or detection system (IDS) probes	Internet

Table 1.2. *Factory perimeters in Industry 4.0*

In addition, Industry 4.0 promotes the application of these technologies to enable distributed communication architectures, such as Peer-to-Peer (P2P), instead of solely relying on standard cloud solutions or other centralized architectures.

The integration of these heterogeneous devices into the industrial cyber environment implies that cybersecurity has to be accounted for as part of the design strategy of companies wishing to implement the Industry 4.0 paradigm. Despite the many improvements brought about by this revolution to the efficiency of manufacturing plants, flaws in cybersecurity can lead to critical impacts on the business model and a loss of competitiveness (Dawson 2018).

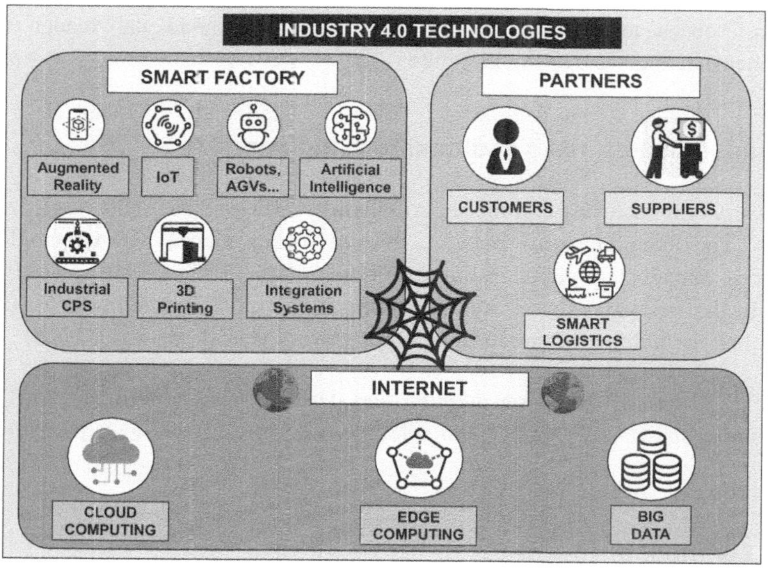

Figure 1.4. *Industry 4.0 technologies*

Several surveys show that only 16% of companies are willing to address the challenges of cybersecurity (Lezzi et al. 2018). The reasons provided include the lack of precise reference standards and the lack of managerial and technical skills to understand and implement them Several organizations working on guidelines and standards help companies understand what schema they should use to reinforce their security and make it compliant. Among these organizations, to name a few, we can find: ANSSI (National Agency for the Security of Information Systems) in France, ENISA (European Union Agency for Cybersecurity) in the European Union and NIST (National Institute of Standards and Technology) in the United States.

A more detailed description of Industry 4.0 concepts and applications is provided in Prinsloo et al. (2019), as well as some recent cybersecurity attacks observed in several manufacturing plants around the world. The authors also propose countermeasures to address a wide range of cybersecurity risks. These risks must first be identified, along with the factors that cause them (Shukla et al. 2022). Another study proposed in Bhamare et al. (2020) focuses on cybersecurity, where the authors analyze in particular the transition from an ICSs in an autonomous factory to a cloud-based environment, resorting to machine learning solutions. A more specific recent survey (Tange et al. 2020) reviewed 262 articles covering all aspects of Industry 4.0 security. In addition to a systematic review of the literature, their main proposals focus on the opportunities offered by fog computing in this field. All of these references are classic surveys that aim to provide the reader with a comprehensive overview of the literature.

1.5. Vulnerabilities, risks and threats in Industry 4.0

Industry 4.0 factories exhibit security vulnerabilities, like most organizational systems. The interconnection between devices makes security more complex and introduces unexpected vulnerabilities (Pereira et al. 2017; Alani and Alloghani 2019). In the industrial sector, manufacturing is the most targeted area by security attacks, and the number of threats is increasing every year (Lezzi et al. 2018).

Section 1.5.1 will focus on defining and describing vulnerabilities, threats and risks. Additional useful content and references can be found in Alqudhaibi et al. (2025).

1.5.1. *Definitions*

1.5.1.1. *Vulnerabilities*

In the computer field, vulnerabilities are defined by Lezzi et al. (2018) as weaknesses that could be exploited by hackers to compromise the system. More specifically, these weaknesses may be found in the system, in security procedures, or in internal controls. They can be classified into three categories which are remote access vulnerabilities, software vulnerabilities and Local Area Network (LAN) or Wireless LAN (WLAN) vulnerabilities.

1.5.1.2. *Threats*

A cybersecurity threat (also known as a cyber threat) is defined by Lezzi et al. (2018) as any circumstance impacting organizational operations, assets, individuals, other organizations or the nation by means of an information system through unauthorized access, destruction, disclosure or modification of information. Several parameters should be considered to analyze a cyber threat:

– the origin of the attack (internal or external);

– the goal;

– the affected cyber layer, including the runtime layer (sensors, actuators), the data transport layer (network) and the application layer (user data storage).

Within the framework of Industry 4.0, the following main cyber threat categories have been identified (Lezzi et al. 2018) to name but a few:

– direct attacks on external access;

– indirect attacks with a service provider to which external access has been provided;

– unknown attack vectors (or "zero-day exploits");

– malware;

– intrusion into neighboring networks.

1.5.1.3. *Risks*

According to Lezzi et al. (2018), a risk is the level of impact on organizational operations, assets or individuals resulting from the potential consequence of a threat and its likelihood of occurrence. In cybersecurity, risks are identified by the loss of certain characteristics such as availability, integrity, confidentiality and authentication.

1.5.1.3.1. Availability

The goal of availability attacks is to render the system unable to perform its usual tasks by overloading it. Their target can be equipment or access to the associated network by disrupting it. The most common types are Distributed Denial-of-Service (DDoS) attacks, which attempt to saturate the bandwidth or other resources in a system, making it unable to respond. Some attacks also affect the network, and more specifically routing operations ("gray hole", "black hole" and relay attacks).

1.5.1.3.2. Integrity

Integrity consists of maintaining data accuracy and completeness. The associated threats are similar to sabotage. They can be used to modify industrial communication protocols or network traffic. A major problem is that most of these protocols are inherited, which underlines the fact that no security considerations were involved during their design.

Some examples of common integrity attacks are "Man-In-the-Middle" attacks, which consists of tampering with and relaying communications between two entities, although these keep on operating as if they were directly connected.

1.5.1.3.3. Confidentiality

Confidentiality threats involve accessing or stealing sensitive data related to industrial processes, configurations, customers or administration. They can be categorized as cyber espionage and carried out in different environments or contexts such as passive network traffic analysis, active code injection into operational applications to obtain security credentials or corruption of control measures.

1.5.1.3.4. Authentication

This includes threats that take advantage of design flaws or software vulnerabilities to obtain privilege elevation and gain access to protected resources. These attacks make use of social engineering techniques such as phishing or spam chains to capture strategic information. Misconfigurations resulting in improper access at both the physical and logical levels can lead to the same security issues.

1.5.2. *Business impacts*

Cybersecurity threats are a serious topic, and they undermine the capabilities of the most advanced companies. A particularly critical year was 2017 as three large-scale attacks were carried out around the world (Kiss et al. 2019):

- "NotPetya" propagated to 65 countries and caused $892 million in damage.
- "Bad Rabbit" targeted critical infrastructure.
- "WannaCry" propagated to 150 countries and caused $8 billion in damage.

As a result, companies infected by these attacks saw their production interrupted for a long time and were not able to efficiently carry out their activities. The impact of these threats is not only technical or financial. It can affect relationships between partners and lead to legal consequences. Table 1.3 describes the analysis found in Corallo et al. (2020) on business impacts filtered by risks. All risks are covered by a range of threats and it is clearly visible that the number of impacts on companies is greater than the number of threats. Each unique threat incurs several business impacts, suggesting that no threat should be underestimated and that policies in terms of cybersecurity investments are needed (Sawik 2020).

1.5.3. *Major cybersecurity threats in Industry 4.0*

The main security threats facing an Industry 4.0 factory can be classified into the following categories (Pereira et al. 2017).

1.5.3.1. *Cyber espionage*

Due to intelligent and connected business processes, Industry 4.0 is vulnerable to cyber espionage. Well-organized cybercriminal groups have turned Industry 4.0 into their favorite playground to steal sensitive information and intellectual property. One such group is the Black Vine group, which primarily targets the aerospace, energy and healthcare industries. The theft of corporate and product data is becoming very widespread, involving in particular software and functionalities that can easily be copied. In Industry 4.0, given that different partners, such as network providers, cooperate between themselves, it becomes easier for these criminals because their attacks can follow many routes and propagate very quickly.

Risks/loss of...	Attacks/threats	Business impacts
Availability	Denial of service DDoS	Productivity loss
		Violation of trade agreements with customers
		Part quality degradation
		Theft of service
Integrity	Critical infrastructure sabotage, machinery or components	Damage to working machinery
		Product quality degradation
		Breach of safety sector standards and regulations
		Violation of trade agreements with customers concerning products
Privacy	Theft of industrial secrets, cyber espionage	Reduction of the business competitive edge
		Damage to the company image and reputation
		Violation of trade agreements with partners concerning data

Table 1.3. *Business impacts related to security threats and filtered by risk (Corallo et al. 2020)*

1.5.3.2. *Denial of service*

DDoS is a cyberattack that aims to make the system unavailable. It can be performed in particular by:

– launching waves of requests on a server to consume all its resources;

– sending malformed input data to crash a process;

– virus infiltration;

– destruction or deactivation of system sensors.

Most devices are interconnected in a factory and, by extension, are interdependent. Consequently, the unavailability of certain devices can be highly critical for a production environment, making these attacks very popular. In addition, with the development of cloud computing, new ways of launching DDoS attacks are emerging, thus driving companies to pay closer attention to them (Haleem et al. 2022).

Unlike cyber espionage in which virtual data is the loss, DDoS attacks have physical impacts with hardware damage such as servers that may need to be replaced (overloading), reconfigured or redesigned.

Another problem with these attacks is that they are unpredictable and very difficult to control.

1.5.3.3. *Supply chain and extended systems*

To improve the efficiency of the supply chain, the Industry 4.0 paradigm proposes the connection between several organizational environments. However, the latter exhibits inherent system vulnerabilities that can be exploited by attackers (Ozarpa et al. 2022).

The source of a vulnerability can be a vendor that is the victim of a phishing attack or credential theft, resulting in massive data exposure for the factory.

1.5.3.4. *Smart security and smart factory*

Most manufacturing companies are not fully aware of the security risks associated with the Industry 4.0 paradigm. They usually mainly deal with safety issues when a serious incident occurs.

Nonetheless, technical products by themselves are not enough to manage these risks (Brandao and Duarte 2022). The human factor is an important point. Employee awareness about safety is also important, from skilled machine operators to secure software and planning engineers. This awareness can be achieved in various ways, such as:

– awareness campaigns involving the entire manufacturing environment;

– research groups in higher education institutions that study cybersecurity topics and provide guidelines for industry professionals.

1.5.3.5. *Advanced persistent threats (APT)*

Threats known as APT belong to a specific class of cyberattacks. They are perpetrated by some groups possessing significant experience and resources. The concept is to take advantage of vulnerabilities to infiltrate the victim's network and remain undetected for a long time (Rubio et al. 2019).

The first APT identified in the industry was Stuxnet in 2010. Stuxnet was designed as a platform to target PLCs and SCADA in order to automate electromechanical processes and cause hardware destruction. It exploited "zero-day" vulnerabilities in the Microsoft Windows operating system and Siemens software. Other examples are Duqu, DragonFly, BlackEnergy and ExPetr. The attack process followed by these APT is generally divided into five steps:

– network reconnaissance to find vulnerabilities;

– communication to initiate the first intrusion by sending "exploits" to the victim. This can be done directly through social engineering (phishing, etc.), or indirectly by compromising a third party such as the vendor;

– tracking of "zero-day" vulnerabilities to execute remote actions using the "backdoors" of the previous steps;

– propagation to other areas of the network to infiltrate new devices in order to gather information or change the behavior of existing hardware;

– filtering the obtained information to transfer it to the attacker's domain.

The first step is possible due to metadata leaks originating in servers, PLCs and sensors. These issues are inherited from the cloud and IoT paradigms (Rudenko et al. 2022) and need to be solved (Rahman et al. 2022).

1.5.4. *Vulnerabilities, risks and threats by perimeter*

In this section, we analyze security flaws in the various predefined perimeters of the factory. Table 1.4 will be used as a support in the following sections.

1.5.4.1. *Manufacturing-production*

Critical equipment in the manufacturing zone can be reduced to ICSss. From a proprietary and isolated architecture, ICSss have become a standard open platform that is highly interconnected with enterprise and public networks (Asghar et al. 2019). New features, including remote access to networks and terminals, have emerged, enabling a wide range of cyberattacks. In addition, these systems are now available on the Internet.

From 1997 to 2015, the number of vulnerabilities increased from 2 to 189 according to the Kaspersky report in 2015 (Asghar et al. 2019). Most vulnerabilities in production are called "zero-day" because developers have just discovered the existence of the flaw while a patch for fixing it has not yet been released.

The usual reasons behind this are highlighted in Lezzi et al. (2018):

– In factories, devices operate for weeks or months without any security updates or antivirus deployment.

– Multiple intrusion routes (carried laptops, USB drives, etc.).

– No existing isolation between the different networks.

The most common threats in this zone aim at compromising availability and integrity, through in particular physical destruction, DDoS attacks, malware and worms, and "zero-day" attacks.

Physical destruction can fall into the category of typical sabotage of industrial equipment. In the manufacturing sector, DDoS attacks mainly focus on routing with relay attacks, selective forwarding, a gray hole, a black hole or botnets that will affect availability.

Malware and worms will slow down operational performance for the purpose of obtaining sensitive information or modifying device behavior and in the end compromising its integrity. Behavior changes can take many forms, such as tampering with communication protocols by exploiting their weaknesses in terms of authentication and data integrity.

Passive traffic analysis can also be utilized to steal confidential information. The method employed may rely on code injection into operational applications with the aim of corrupting control mechanisms, hacking the end user and then accessing the data.

Privilege elevation can be achieved by taking advantage of security vulnerabilities existing in software. In 2015, IBM X-Force research reported that 45% of all attacks focused on unauthorized access.

The easy mobility of operators in the factory and their many interactions with mobile devices (laptops, smartphones, tablets) increase the risks associated with these threats.

Applications and device configurations and access controls need to be rigorously verified in this zone.

1.5.4.2. *Logistics*

In the logistics field, many wireless devices exhibit certain vulnerabilities, such as (Kondiloglu et al. 2017):

- Wi-Fi networks, especially when they are not encrypted;
- business applications that still use the HTTP protocol;
- installation of malicious applications.

Network vulnerabilities (unencrypted Wi-Fi, HTTP, etc.) expose almost all information sent by devices to hackers. For mobile terminals, all data scanned with the barcode scanner such as part references, serial numbers, product destinations as well as information about the IT infrastructure such as servers and databases are involved.

The installation of malicious applications could be the source of malware attacks. As an example, "HummingBad" and "HummingWale" affect Android devices and deploy personal data collection applications that are sold in the process.

1.5.4.3. *Active supervision*

The active supervision zone belongs to the company network and mainly includes office equipment. Some possible flaws in this zone are (Kondiloglu et al. 2017; Asghar et al. 2019):

– the lack of antivirus software or out-of-date signatures that could infect the entire industrial control system (ICS) via the Internet and make it unavailable;

– people unaware about security who click on malicious links;

– computers left unlocked when leaving the office;

– access to unauthorized files/websites/data;

– connection to external devices;

– installation of unlicensed (pirated) programs.

These vulnerabilities could be the source of several threats such as:

– social engineering attacks (phishing);

– network attacks;

– viruses, malware, worms and ransomware;

– theft, loss or hardware and data failure;

– data transfer to and from unauthorized devices.

1.5.4.4. *Research and development*

This zone contains innovative equipment that can become a source of potential threats due to a lack of perspective on the technologies used. This zone can also be targeted by attacks because of the confidential and valuable information that attackers might find.

1.5.4.5. *Living zone*

In the living zone, VoIP phones are the main equipment identified. While VoIP allows businesses to save on network costs, it also brings about new threats and security risks (Omari et al. 2022). Some of these risks are common, such as DDoS attacks that overwhelm the VoIP server with SIP (Session Initiation Protocol) call signaling messages. Such attacks are dangerous because they do not require penetration of the entire network.

Viruses and malware can also affect VoIP phones because VoIP configurations make use of "softphones". With VoIP networks, mobile malware is also a problem as many users make VoIP calls with their smartphones. This means that once the malware has infiltrated the smartphone, it can access and steal valuable information.

Another threat is *vishing*, which is the vocal counterpart of email phishing. Employees, suppliers and customers are required to share sensitive information.

Phreaking occurs when a hacker gains access to a company VoIP network and uses it to steal corporate data, change the network plan, or make expensive calls, resulting in huge costs for service providers.

Eavesdropping is also a common cybersecurity threat that is very difficult to prevent. Hackers are able to access and listen to VoIP calls by capturing unencrypted VoIP traffic. In this way, they are capable of performing identity theft, as well as theft of VoIP services.

The latest threat is *Spam Over Internet Technology* (SPIT). For hackers, this consists of capturing thousands of VoIP IP addresses and then sending voice messages to a VoIP system.

1.5.4.6. *External zones*

Outdoor spaces are divided into two categories which are physical and virtual.

1.5.4.6.1. Physical external zone

The physical external zone consists of IP surveillance cameras that fall under the umbrella of the IoT. Threats to these types of devices include:

– interference with the routing protocol mode of operation by way of jamming and interference in order to disrupt communications;

– resource exhaustion through the use of vulnerabilities in software that controls devices or with malicious code (*malware*);

– manipulation of routing information to influence traffic, as in a Sybil attack (Rubio et al. 2019). These attacks are gateways to others, such as black hole or denial of service;

– side-channel attacks to expose device information (battery, memory) or routing and topology information to identify vulnerable devices in the infrastructure;

– injection of fake nodes that can execute code or inject illegitimate traffic in order to control large areas of the network or to eavesdrop;

– poor access control resulting in unauthorized access to protected resources.

1.5.4.6.2. Virtual external zone

The second category is the virtual external space, which encompasses everything related to cloud computing. The latter is attractive due to its low investment cost and how easily companies can deploy it. Many organizations resort to cloud computing for storing their data, but also for hosting certain processes. It can even be used on the IoT to acquire sensor data, but also as a way for customers to manufacture a product through a shared network of suppliers throughout its lifecycle. These innovations introduce the following threats:

– DDoS attacks that exploit vulnerabilities inside the scheduler component of some hypervisors to charge for the service and make it unavailable (Rubio et al. 2019);

– malware injection to replace a legitimate cloud instance service such as a virtual machine with a malicious service to access the exchanged data;

– side-channel attacks that force machines to study electromagnetic radiations and access their resources;

– shared memory attacks that analyze the cache or main memory to obtain technical information about the infrastructure, running processes or to access the memory dump of virtual machines;

– social engineering attacks that capture customer information from different applications. The objective is to obtain sensitive data such as accounts and passwords in order to host malicious services in the cloud.

Most of these attacks are referred to as advanced persistent threats (APT), which can be executed primarily by experienced, resourceful attackers. It should be noted that here we have not considered data sovereignty-related risks and geopolitical risks, when cloud services do not fall under the legislation of the country in which the factory is located.

1.6. Review of cybersecurity solutions

This section will provide a detailed examination of some of the existing solutions against cybersecurity threats. These solutions have been selected in order to present the main trends concerning conventional and innovative solutions that can be applied in the different perimeters defined above.

The following two notions are necessary for a better understanding of the solutions presented: countermeasures and long-term solutions found in the literature.

1.6.1. *Definitions*

1.6.1.1. *Countermeasures*

Countermeasures can be seen as the short-term defense mechanism against a threat or attack. They represent a set of actions and techniques to eliminate a threat or prevent it in order to limit the damage it could cause. They can also help flag the threat so that corrective action can be taken (Lezzi et al. 2018).

1.6.1.2. *Solutions*

Solutions are the long-term way to deal with threats. They are often defined with different names such as approach, methodology or architecture. They represent a comprehensive set of actions for protection against different types of cyberattacks. These solutions can be classified into three categories (Asghar et al. 2019): security assessment tools, intrusion detection and prevention technologies as well as ICSs risk management.

	Possible threats in perimeters					
	Threats	Manufacturing-production	Logistics	Active supervision	Living areas	External areas
Availability	Device hijacking	X				
	DDoS attacks	X	X	X	X	X
	Path attacks	X	X	X		
	Node resource depletion	X	X	X		
	Service theft	X	X			X
Integrity	Incorrect configuration	X	X			X
	Malware injection	X	X	X	X	X
	Fake data injection	X	X			
	Identity theft	X		X	X	
	Routing information tampering	X	X			
Confidentiality	Sensitive data theft	X		X		X
	Node state exposure	X	X	X	X	X
	Passive traffic analysis	X	X	X	X	X
	Infrastructure-related information exposure	X				X
Authentication	Privilege elevation	X		X		X
	Social engineering	X		X	X	X
	Inefficient access control	X	X	X		X
	Node identity theft	X	X			

Table 1.4. *Risks and threats by perimeter in Industry 4.0*

Security assessment tools are able to provide secure experimentation with realistic test scenarios (attacks, infections, etc.) in order to detect security issues before initiating production.

Intrusion detection and prevention technologies highlight approaches to securing ICSss by introducing new components or upgrading existing architectures. For the management of ICSs risks, guidelines, standards and measures are proposed to ensure that the security implemented is protected against developing threats.

1.6.2. *Cybersecurity countermeasures*

According to Lezzi et al. (2018), the following three-tiered approach can be used to ensure the security of ICSs:

– perimeter hardening by isolating the factory network from the office network using firewalls and a demilitarized zone (DMZ);

– network multi-layered defense to contain attacks;

– user isolation within a separate zone/network.

Countermeasures	Protects against/prevents
Communication encryption	Data corruption Information disclosure
Stored data encryption (signature)	Repudiation attacks
Data stream encryption	Data tampering Repudiation attacks DDoS threats
Firewall/gateway and proxy	Unwanted network access (from/to)
Access control/multiple authorization	Unauthorized access (information, physical areas)
Software updates	Vulnerabilities (system, software, business, firmware)
Secure communications (VPN, Wi-Fi and IP)	Eavesdropping Unauthorized connection Data analysis
Antivirus and anti-malware	Malware

Table 1.5. *Countermeasures for cybersecurity in Industry 4.0 (Lezzi et al. 2018)*

These measures enable protection against unwanted access to and from the Internet, but are also used for keeping services and zones separated between plant systems. Encryption is the most popular of these measures and is available at several levels:

– communication encryption to prevent information tampering and disclosure;

– encryption of stored data to prevent repudiation attacks;

– data stream encryption to avoid all previous problems.

To be effective, all countermeasures must be continuously updated, including encryption protocols, firewall signatures, security controls (patches), and monitoring (log analysis). Table 1.5 describes the most popular countermeasures summarized from Lezzi et al. (2018) and how they can help prevent cyber threats.

1.6.3. *Proposals for cybersecurity solutions in the literature*

This section will highlight and present in detail some of the cybersecurity solutions currently available in the literature (Table 1.6). For each solution, we describe its purpose and design structure, as well as an evaluation of how it has contributed to improving cybersecurity.

1.6.3.1. *Protection of physical environments*

1.6.3.1.1. Background and goal

Modern businesses need both software and physical solutions for the security of their information. Despite the effectiveness of software controls, these remain virtual and there is always the possibility that an interface may escape the control of the system and compromise it. This interface can be human, and the flaw appears when information is verbally exchanged. In addition, the explosive growth of wireless networks in information system architectures has further increased the possibility of information leakage (Illési et al. 2018). For these reasons, Kiss et al. (2019) proposed the creation of a physical information security mechanism. This solution is implemented in the form of "protected zones" (Figure 1.5), which designate spaces where sensitive and valuable data can be securely exchanged, both acoustically and visually.

1.6.3.1.2. Design and approach

This solution aims at the following goals:

– prevent access to unauthorized persons;

– restrict information propagation to the limits of protected zones;

– maintain a homogeneously protected and controlled environment.

To achieve these objectives, three layers of solutions are proposed.

The first layer consists of the conventional mechanisms used to defend protected installations: admission authentication, people screening, definition of authorizations and security systems.

With the first layer, we can be sure that the people present in the zone are all authorized, however there is always a risk because the information can still be "eavesdropped" by someone outside.

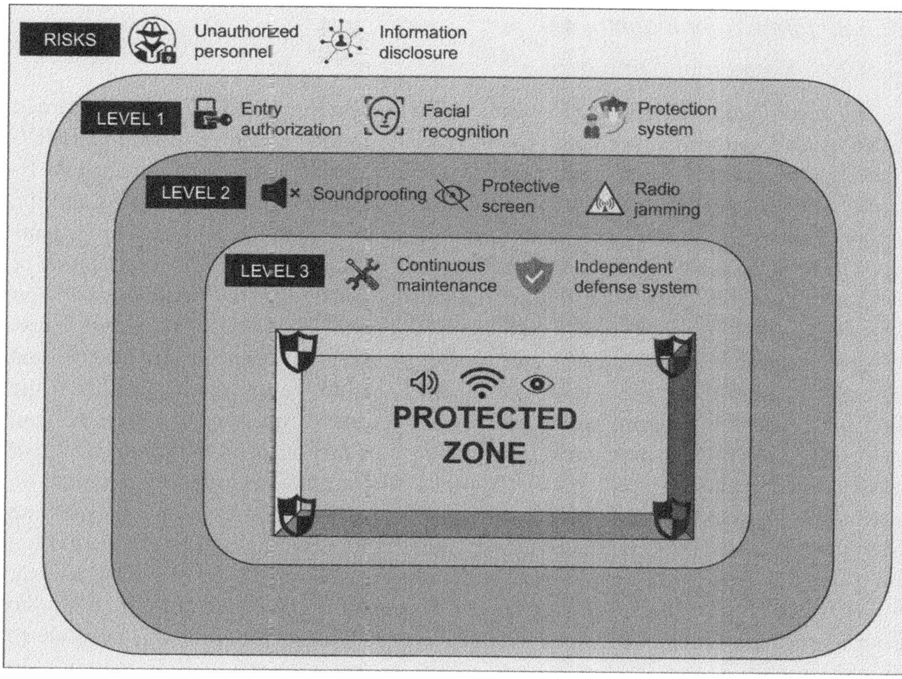

Figure 1.5. *Protection of physical zones*

The second layer takes care of the prevention of information transmission outside the protected zone by physical characteristics (visual, sound, radio) through the use of acoustic shielding, visual shielding and radio shielding.

This security proposal defines the last layer as being responsible for maintenance, which involves continuous maintenance and an independent defense system.

The idea underlying maintenance is to preserve the homogeneity of the environment, search for alien objects and monitor radio signals. An autonomous safeguarding and surveillance system has also been recommended to complete the security of these protected zones.

1.6.3.1.3. Results and highlights

The author suggested that although new technologies introduce new threats, the preparation and creation of these controlled zones jointly with other actions could result in preserving integrity, confidentiality and availability in the long term.

1.6.3.2. *Firewall for Industry 4.0*

1.6.3.2.1. Background and goal

The manufacturing execution system (MES) is the intermediate system between the ICSs and enterprise applications such as enterprise resource planning (ERP) software. It improves the transparency of manufacturing data. Sensor data can be used to compute real-time performance indicators or monitor machine status and the quality of manufacturing processes. Improved interconnectivity of IT systems exposes devices such as PLCs to cyberattacks, which could disrupt production or infect other systems. The most widespread attacks concerning the MES are based on network scanning/probing where defense-in-depth is an efficient countermeasure. Cybersecurity standards typically proposed network architectures split into several segments including firewalls in between them to minimize security risks. Given the need for defining configuration rules in a flexible and secure way, software-defined networking (SDN) is probably a key technology for this purpose (Nunes et al. 2014; Satasiya and Raviya Rupal 2016). Another recent related proposal focused on network performance is presented in Zeng et al. (2019). Software-defined networking (SDN) is a technology capable of network alternation, releasing critical information and providing new services to run on-demand applications. It provides a clear overview of the network architecture to administrators and allows users to programmatically control the network architecture. This means that it is possible to modify network access on demand and minimize the exposure of ICSs networks to attackers. On this topic, Tsuchiya et al. (2018) proposed a protection network structure based on an SDN firewall specifically designed for industrial networks, without jeopardizing network flexibility (Figure 1.6).

1.6.3.2.2. Design and approach

The proposed solution has three objectives:

1) the creation of segments without reconfiguring existing networks, while using a DMZ (demilitarized zone) vertical integration;

2) the development of one-way access mechanisms (server access only when a client requires a connection);

3) the reduction of flaws in access rules due to the frequent addition or replacement of devices in certain zones.

The two main functions of the SDN firewall are packet filtering based on automatically applying access rules to a group of devices as well as operating as a gateway between network interfaces to avoid any changes to the existing network configuration.

This packet filtering function means that only whitelisted applications are authorized to access industrial control devices. The whitelist is managed by network administrators, whose management task is greatly simplified, because it is the only thing that needs to be done to comply with the security standards of ICSss. At the component level, the firewall contains:

– the transparent firewall that applies access rules and implements an access control system based on the OPC UA standard;

– the firewall controller that stores the production system configuration and manages the rules.

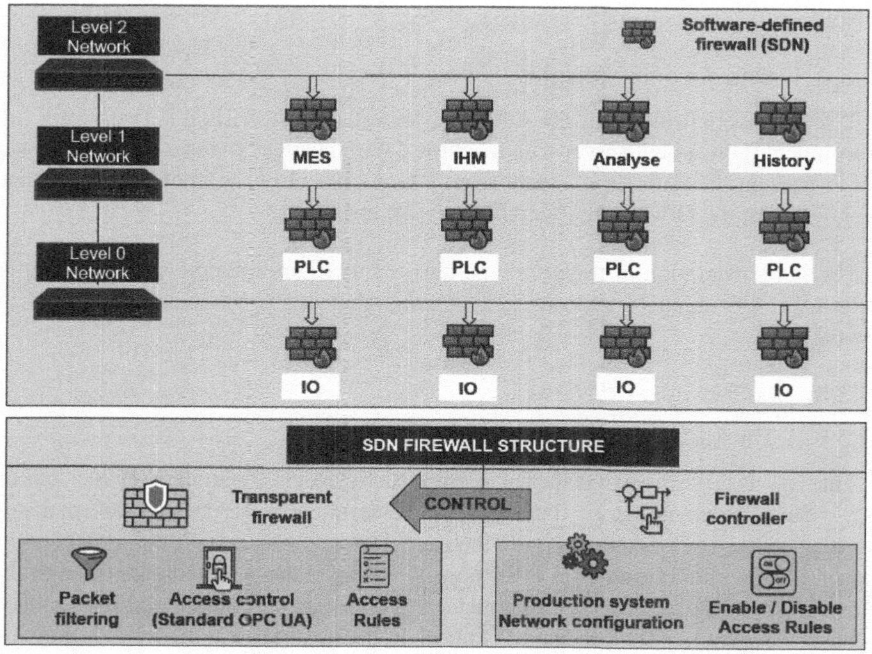

Figure 1.6. *Software-defined firewall (SDN)*

The Open Platform Communication Unified Architecture (OPC UA) standard used for access control in the transparent firewall is a machine-to-machine communication protocol for industrial automation.

This is an improved development of the original OPC protocol to satisfy the emerging requirements of industrial automation in Industry 4.0. Its most significant innovations are:

– cross-platform implementation;

– scalability (smart sensors, smart actuators, etc.);

– multi-thread or single-thread;

– security enhancement (new standards);

– configurable waiting times;

– chunking of large datagrams.

Security in OPC UA consists of authentication, authorization, encryption and data integrity relying on signatures. In addition, the communication stack makes use of firewall-compatible transmissions, which is why this standard was used to design the SDN firewall.

1.6.3.2.3. Results and highlights

The author confirmed that the solution was tested in the virtual network of a full environment (OPC client/server with machine data exchange) and that the firewall was able to prevent the security scanners from taking control of the application port and other details at the OPC server operating system level.

The implementation of the prototype allowed us to improve the security features of the OPC UA standard and provided a comprehensive security solution for ICSs networks.

1.6.3.3. *Security in IoT devices*

1.6.3.3.1. Background and goal

The IoT is a term that was first employed by Kevin Ashton in 1999 (Ashton 2009). There is no universal definition of the term, however it suggests that any everyday object can be equipped with the capability to measure anything related to its environment, communicate with other objects and send the detected data through the Internet (Whitmore et al. 2014). In a factory, it connects several technologies related to sensor development and machine control. In the context of Industry 4.0, they are becoming popular due to the interconnection between data from industrial workshops and the possibility of providing feedback about system operation. Their use leads to the new concept of CPSs, which are associated with the implementation of the IoT, and involve the use of sensors for collecting, processing and utilizing data in the cybernetic world. In terms of IoT hardware connectivity, there are emerging networks called Low Power Wide Area Network (LPWAN), such as Sigfox, LoRa and Weightless (Ali et al. 2017). These networks are specifically designed for the IoT because they are low power consumption protocols. Furthermore, they enable two-way communication between objects and the outside world using the LPWAN network and the network provider's cloud for internet connectivity.

The main underlying drawback that hampers a wider acceptance is their weaknesses in terms of cybersecurity (Atzori et al. 2016), which are caused in particular by the heterogeneous connectivity and the threats deriving thereof (breach of privacy, etc.), and can lead to serious consequences for IoT technology users (Lin et al. 2017). According to Wolf and Serpanos (2018), these systems must be designed and operated following a unified view of safety and security features (Miraz et al. 2015). In the context of the smart factory, one of these threats could be the result of weaknesses linked to the use of cloud computing (Hassija et al. 2019). With these considerations in mind, an ontology-based cybersecurity framework for the IoT was proposed by Mozzaquatro et al. (2018).

1.6.3.3.2. Design and approach

The proposed framework focuses on:

– business-side monitoring;

– security analysis and classification in a knowledge base;

– the design of security services;

– improved security mechanisms for business processes and technology assets.

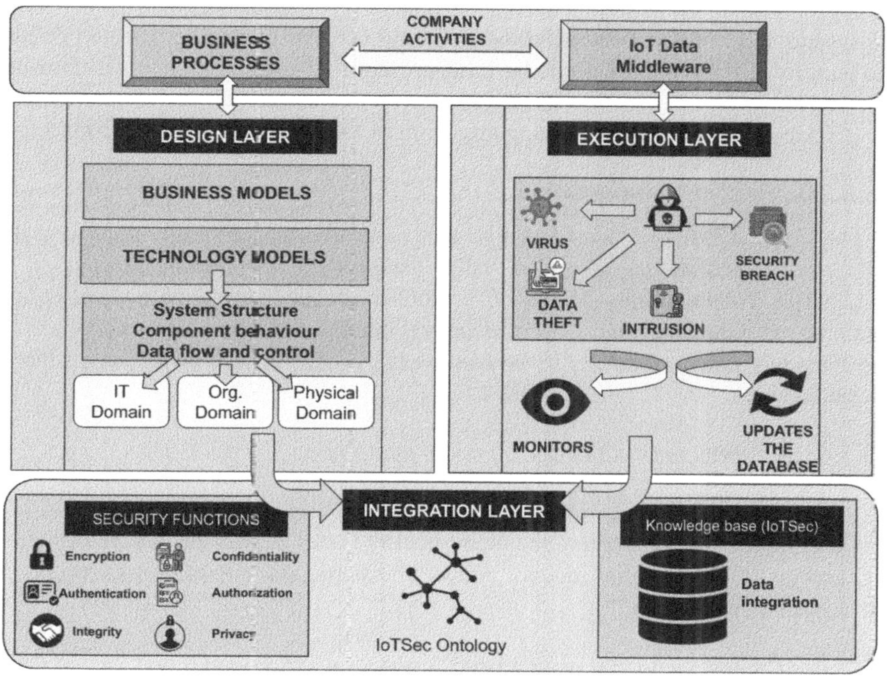

Figure 1.7. *Ontology-based framework for IoT cybersecurity*

Figure 1.7 is a representation of the framework architecture. It is broken down into three layers: two layers that deal with cybersecurity at design and run time, and the integration layer used during both stages. The concept behind the design layer is as follows: it is assumed that a company needs to implement specific services at all times in addition to those already in use, which requires adaption in order to meet the device constraints.

The Model-Driven Service Engineering Architecture (MDSEA) is a software development methodology that focuses on creating and making use of conceptual models (for the framework: business models, technology-independent models and specific models) related to a specific problem. Using this methodology, the design layer is capable of generating code from a high level of abstraction to accelerate service design and adaptation, as well as deployment time. Managers can then collaborate with developers and contribute to building new features.

The runtime layer has two goals: to monitor and update the knowledge base. Monitoring is the process of detecting intrusions, data theft, viruses and any other attempt to exploit security vulnerabilities. All these situations are analyzed to identify the appropriate solutions among the IoTSec pool of security services in order to restore the system and improve cybersecurity. Updating consists of adding detected threats and security analysis to the knowledge base, preventing these threats from reoccurring. The data integration layer provides information on threats and vulnerabilities based on the IoTSec ontology, which is an ongoing work of Mozzaquatro et al. (2015).

1.6.3.3.3. Results and highlights

Finally, the authors consider that the cybersecurity ontology improves the efficiency of security operations and helps analysts extract relevant information to characterize vulnerabilities. However, a number of questions remain open before obtaining a multi-layered cybersecurity intelligence ontology capable of understanding potential threats in the context of cybersecurity, which is a constantly developing area.

1.6.3.4. *Authentication of cyber-physical systems*

1.6.3.4.1. Background and goal

Long value chains are one of the main security issues of Industry 4.0. Information Technology (IT) paradigms do not reflect the particular circumstances found in operational technology (OT).

Data of many kinds are accumulated in the production zone, and are used for quality control and predictive maintenance. However, only some of them are critical to protection. An OT solution should focus on nomenclature, design information and control parameters (Wegner et al. 2017).

In the manufacturing environment, the architecture currently in use is in general entirely separate from production environments, with isolated manufacturing devices.

There is no control over contractors in multi-stage production when data are no longer on the server.

Similarly, operators are not supervised once the data are received and many issues arise, such as data corruption. As established by Pan et al. (2017), the threats identified for additive manufacturing apply to digital manufacturing devices of all shapes and sizes. Consequently, a new paradigm is proposed by Wegner et al. (2017). It aims to limit and protect the flow of information in devices used during the outsourced stages to enhance perimeter security.

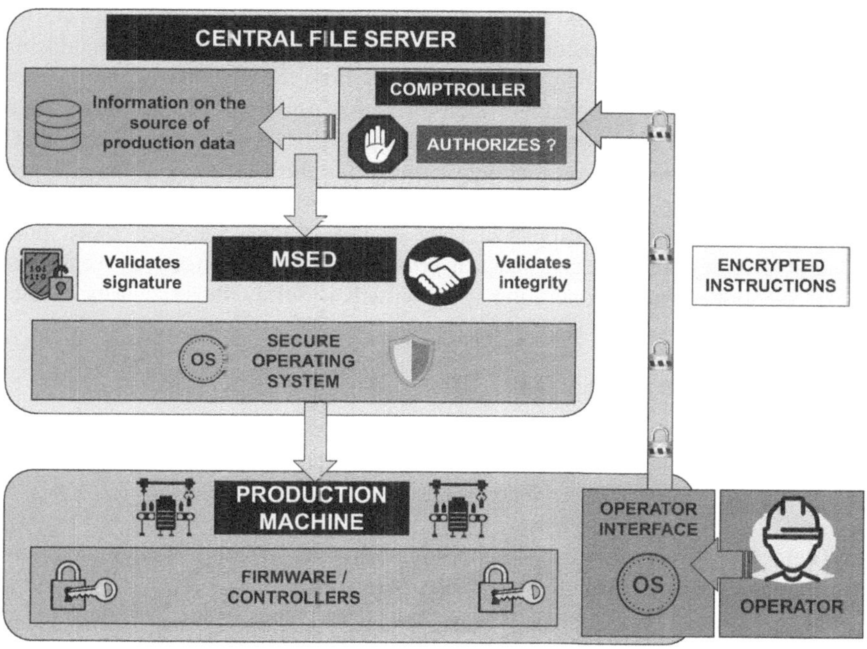

Figure 1.8. *Direct-to-machine approach for the protection of manufacturing cyber-physical systems*

1.6.3.4.2. Design and approach

This new manufacturing information architecture is based on the holistic approach according to which data must be sent directly to the relevant manufacturing device. This approach is also called direct-to-machine communication (Figure 1.8). The fundamental security issue underlying it concerns authentication and authorization:

– Is the request sent to the device authentic?

– Is the sender allowed to send this request?

To solve this problem, the proposed architecture comprises two components. The first one is encryption with a manufacturing security enforcement device (MSED). It relies on asymmetric cryptography. The public key is used to verify that actors or entities have the proper private key, thereby authenticating them.

The second is a software program called "comptroller" that runs on a production network and authorizes every action taken. It takes the input data, provides a key and stores the output data. Output data are appended at the end of a virtual document, which is the provenance record of a produced part. The data transmitted between the "comptroller" and the production device will then be processed by a MSED whose exhaustive description can be found in Graham et al. (2016). The MSED is located upstream of the manufacturing equipment and authenticates manufacturing instructions issuing from the cloud by verifying both the integrity of the instruction data, as well as the "comptroller" identity and permissions. To this end, the best method consists of using encryption and unique data signatures to ensure that the data source is authentic. Another requirement for the MSED is to have a secure operating system (OS). Common operating systems for embedded systems are Microsoft Windows™ or Linux™. They provide a wide range of tools, but are vulnerable to zero-day attacks that can prove devastatingly effective. Wegner et al. (2017) mentioned smaller operating systems such as SeL4, whose security have been formally verified.

1.6.3.4.3. Results and highlights

An MSED device prototype using the SeL4 microkernel has been developed in collaboration with True Secure SCADA, LLC and has been confirmed to be compatible for general industrial control as well as manufacturing.

There are strong arguments supporting direct-to-machine communication for overcoming cyber-physical security challenges, namely:

– its characteristics (operator authorization, monitoring and control, device support, distributed responsibility, location independent);

– its contrast with existing solutions (built-in security, layered-based encryption, always up-to-date, integrity protection, minimized sharing);

– its role of building a responsive production environment (detailed monitoring with the controller, collaboration, distributed manufacturing, automation).

One of the challenges concerning the adoption of this approach is the willingness to manage a higher security level at the cyber-physical scale, or that IT services acknowledge production hardware as another digital component to be included in the list of elements to be protected. In Industry 4.0, these devices are no longer separated from the data flow, and the direct-to-machine approach acknowledges this fact.

1.6.3.5. *Predictive cyberattack detection*

1.6.3.5.1. Background and goal

Conventional cybersecurity architectures focus on mechanisms that guarantee confidentiality, authentication, integrity, access control and non-repudiation to prevent network intrusions and attacks. Nonetheless, the existing security landscape is characterized by large, fast, persistent and highly sophisticated continuously changing attacks.

For critical systems belonging to Industry 4.0, autonomous detection and response to cyberattacks is required in order to achieve robust cybersecurity with defense-in-depth. A cyberattack detection algorithm has been proposed by Thames and Schaefer (2017) to defend Industry 4.0 systems, as well as other internet-based systems. This algorithm is based on ensemble intelligence with neural networks to utilize a classification output providing feedback to active response mechanisms. The underlying goal is to show how computational intelligence-based approaches can be employed in Industry 4.0 cybersecurity.

1.6.3.5.2. Design and approach

Usually, cyberattack detection systems require algorithms that collect and analyze data generated by various events occurring in a cyber environment. One of their main problems is lack of accuracy, and inaccurate results can negatively impact system performance and lead to security issues (false reports, unnoticed intrusions, etc.).

Computational intelligence systems (CISs) are adaptive systems endowed with decision-making capabilities, and they are specifically designed to manage large volumes of data (with noise) in their decision-making process. Therefore, they appear to be a logical choice when designing new algorithms for detection systems. These systems use technologies such as *Machine Learning* and *Deep Learning* that are able to accurately discover the essential differences between normal and abnormal data (Zhao et al. 2017; Liu and Lang 2019).

The proposed classification algorithm is called Neural Network Oracle (NNO) and includes three components: neural networks, genetic algorithm and the NNO.

Figure 1.9 is a representation of the ensemble intelligence framework that uses the NNO classification algorithm for predictive cyberattack detection in advanced manufacturing.

Neural networks, also known as artificial neural networks (ANNs), loosely mimic biological processes and are used to solve AI problems. One of their most interesting characteristics is training-based self-learning with datasets. In the NNO, a collection of neural networks will train with a first set of audit data. The result of this first classification will then be sent to the NNO.

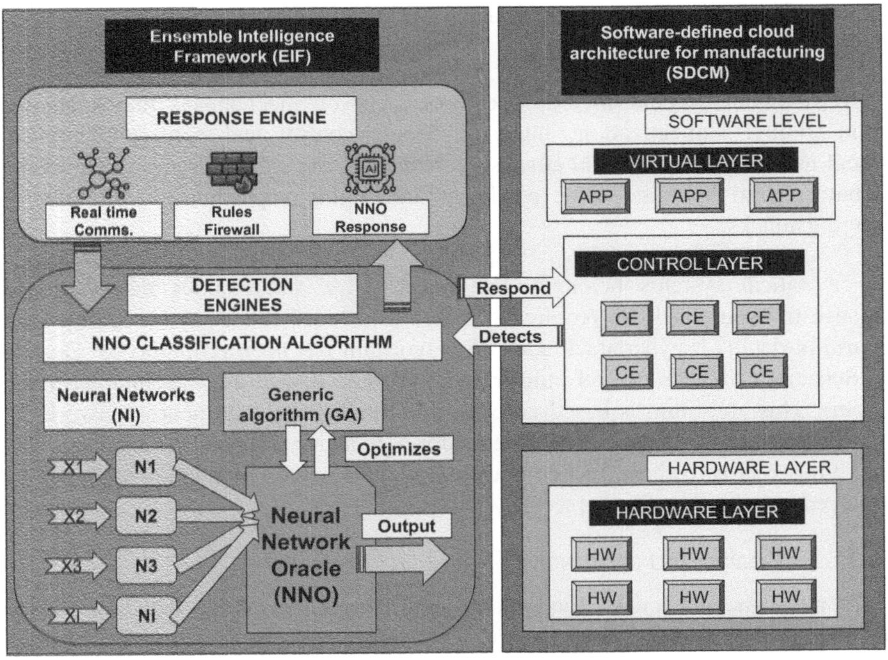

Figure 1.9. *Predictive cyberattack detection with ensemble intelligence in advanced manufacturing*

The genetic algorithm (GA) is based on the process of biological evolution and the ability to adapt over time to changing environments. In the NNO algorithm, the GA is responsible for finding the most optimal parameters in order to reduce the error rate and increase accuracy. To obtain these optimal parameters, a fitness function evaluated with the neural network responses is used.

The neural network oracle is trained with a secondary dataset consisting of the outputs from the previous neural networks and the original dataset. To minimize errors, it utilizes the optimal parameters provided by the genetic algorithm.

1.6.3.5.3. Results and highlights

The author integrated the NNO classification-based ensemble intelligence framework into a software-defined cloud architecture for manufacturing (SDCM) (Thames and Schaefer 2017). The SDCM is divided into three layers: virtual, control and distributed hardware.

Given that the control layer is the one with the best knowledge of activities and communications, it will be employed as a data collection point. The control layer will continuously feed the ensemble intelligence framework (EIF) with data, and the EIF will be responsible for analyzing the detected data and solving the detected anomalies.

In terms of performance, the NNO was trained and tested with the *CUP99* intrusion detection dataset, and achieved a good classification performance. It was concluded that it could be coupled with active response mechanisms within the framework of Industry 4.0 to counter cyberattacks.

1.6.3.6. *Predictive maintenance system protection*

1.6.3.6.1. Background and goal

ICSs have been increasingly targeted by hackers since the beginning of the 21st century due to the potentially significant damage they could inflict on the system and its environment if the attacks are successfully executed. The ICS network can be broken down into three levels (McLaughlin et al. 2016; Sicard et al. 2018):

– level 0: the operational part with sensors/actuators;

– level 1: the control part with PLCs/HMIs;

– level 2: supervision with control room/SCADA.

IT solutions such as firewall/DMZ are typically used at level 2 and above for protection against DDoS or Man in the Middle (MITM) attacks. Solutions in this layer work well because they are very similar to traditional IT infrastructure.

However, these solutions do not work in real-time layers 0 and 1, which have their own inherent attacks (random attacks, fake data injection). Four types of attacks are considered: namely direct, sequential, time-based and over-solicited attacks.

An innovative methodology was proposed by Sicard et al. (2019) based on the concept of "automation knowledge". This is an in-depth study of Sicard et al. (2018) by the same author. The solution will focus on the protection of low-level elements (level 0-1) such as PLCs, computer-integrated manufacturing (CIM) architecture sensors, taking into account safety and security aspects. The final objective is to be able to detect malicious commands issued by the PLC.

This approach can be seen as a last shield to protect the system from cyberattacks assuming that hackers have already broken through the previous defense levels.

1.6.3.6.2. Design and approach: models and filters

The first step is based on behavioral patterns (Mitchell and Chen 2013), which represent the normal operation of the system. It is divided into four parts:

– risk assessment to determine the critical zones to be protected (prerequisites);

– parameter identification (offline);

– control of the filter model generation (offline);

– execution of detection mechanisms during operation (online).

Risk assessment is a prerequisite. The methodology assumes that a risk analysis has been performed with the system. This analysis should highlight the events that could be harmful to the system and infer which part of the ICSs should be protected first. To this end, the input/output (I/O) list of the PLC has to be modeled.

Parameter identification consists of creating the states from the I/O list created by the risk assessment. This will help define both the scope and the states used in the behavioral models. Mathematically, these states are defined with combinatorial constraints based on the values of the sensors and actuators. The following types of states are accounted for in an ICS:

– optimal (compliance with the control law as well as the physical constraints of the system);

– dangerous (compliance with the physical constraints of the system only);

– forbidden (damage to the physical system);

– accessible (all admissible states);

– inaccessible (all states to avoid).

Next, time constraints must be determined to identify time-based attacks. These time specifications introduce another dimension to the characterization of the system behavior as well as another protection level for the ICSs. The use of combinatorial and time constraints is required to ensure the efficiency of detection algorithms.

The generation of the control filter model is intended to represent the industrial system with the process and control model. The operation stage corresponds to the availability of the control and report filters in the ICSs. These filters should be located just after the sensors to ensure command integrity and to detect any malicious behavior (Figure 1.10).

The next question consists of determining the desired protection level with these filters. Three levels of security are taken into account:

– goods and people security: guarantee a set of states according to which the system can develop without danger;

– quality: guarantee the proper execution of the control law and monitor the system;

– equipment protection: monitor actuator loads under overly intense or too frequent commands (in order to reduce wear, risk of failure and equipment maintenance).

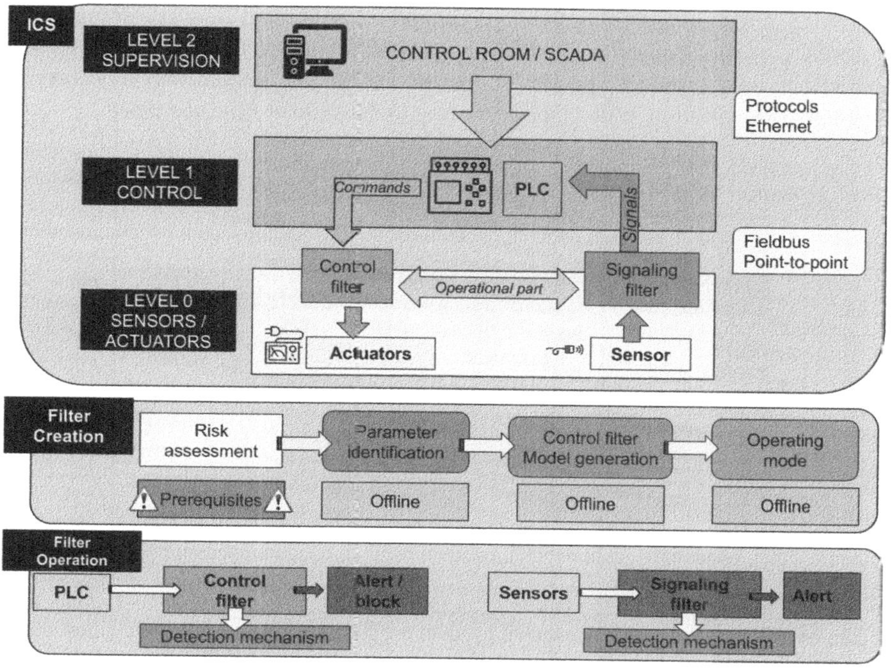

Figure 1.10. *Filter implementation in the behavioral model approach for the protection of industrial systems*

1.6.3.6.3. Design and approach: detection based on the notion of distance and trajectory

The filters described in the previous section allow the implementation of rules, but they can only block attacks when the critical state is about to be reached. A mechanism for detecting deviations from normal behavior is necessary to implement an exhaustive methodology. This mechanism is based on three notions, which are distance, the shortest path to the critical state and trajectory. Figure 1.11 is a simplified representation of the proposed detection mechanisms.

The notion of "distance" is linked to the ICSs states and was introduced in Carcano et al. (2011). It represents the deviation between the current state and a set of critical states, all of which are states that the system should not reach. The benefit of this concept resides in the fact that it provides operators with indications about proximity to the critical zone. The "shortest path to critical state" is the smallest number of commands that must be applied before reaching a critical state. The overall goal is to calculate, for each achievable state, the closest distance to a critical state. These computations are performed offline and integrated into the filters to perform the detection algorithm, while taking into account real-time constraints. The last notion is that of "trajectory", which is complementary to the notion of distance. The distance only provides specific information about the system, but trajectory is defined as "the evolution of distance according to the state or time sequence".

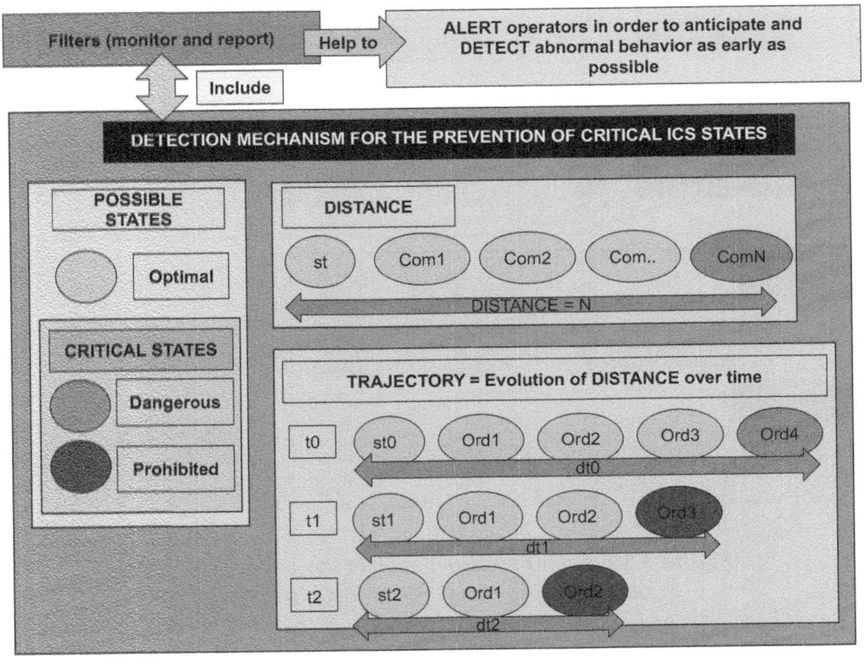

Figure 1.11. *Distance and trajectory-based critical state detection mechanism for the protection of industrial systems*

The following detection mechanisms can be implemented with the distance/trajectory approach:

- context detection;
- anomaly detection (combinatorial and temporal);
- equipment degradation.

The main advantage of the proposed mechanisms is that operators receive warnings, deviations are anticipated and abnormal behavior detected at its earliest stage.

1.6.3.6.4. Results and highlights

The author confirmed that the approach showed good results for detecting cyberattacks that affect physical systems. The particular improvement could be seen in blocking commands before driving the system into a critical state, and in detecting sequential attacks using combinatorial and time constraints. However, some improvements were proposed, such as:

– using a safety/security approach to improve the effectiveness and consequently the number of attacks accounted for;

– considering the system behavior between two stable states.

1.6.4. Honeypots *and* digital twins

1.6.4.1. *Overview*

Several innovative solutions can be found in the literature, namely the cybersecurity immune system for Industry 4.0 (Petrenko 2020), to name just one. However, in this book, we will focus on two popular approaches: *honeypots* and *digital twins*. Honeypot systems are passive monitoring systems capable of early alerts in production environments and critical infrastructure. Their main feature is to issue alerts if the infrastructure has been penetrated by hackers or malware-related activities.

Technically, a *honeypot* consists of data that appears to be a legitimate system component presenting valuable resources for attackers. However, in fact, the *honeypot* is isolated and monitored to block or analyze attackers. In more specific terms, the idea is to bait attackers. The main advantages of this solution are:

– awareness that someone or something is trying to exploit your company's critical systems;

– the attacker loses valuable time attacking a decoy system, instead of the real infrastructure;

– more time is thus given to the security team to block the attack;

– the company will not waste time with false positive alerts because there is no reason for any communication to take place from or to the *honeypot* system.

1.6.4.2. *Honeypot-based solutions*

Existing honeypot-based solutions are usually distributed systems capable of collecting and analyzing threat or attack-related information (Rubio et al. 2019). The purpose of this analysis is to determine the type of attack, the existence of infected devices, as well as the activities carried out with the system.

Cybersecurity solutions	Concerned equipment	Protects from...	Concerned factory perimeter(s)
Physical environment protection	All	Unauthorized physical access, information spreading (acoustic, visual, radio)	All
Software-defined firewall	Manufacturing Execution System (MES)	Network scanning	Manufacturing-production
IoT ontology framework	IoT devices	External attacks (viruses, data theft, security breaches)	Manufacturing-production/logistics
Cyber-physical system direct authentication	Cyber-physical systems (CPSs)	Unauthorized instructions to production machinery	Manufacturing-production
Advanced manufacturing set learning framework	Industrial control systems (ICSs)	Predictive cyberattack detection using neural networks	Manufacturing-production
ICSs predictive maintenance by establishing behavioral models	Industrial control systems (ICSs)	Predictive cyberattack detection using the notion of distance and critical states (sabotage detection, for instance)	Manufacturing-production

Table 1.6. *State of the art of cybersecurity solutions in Industry 4.0*

1.6.4.2.1. Attica Networks ThreatMatrix

ThreatMatrix is Attica Networks' leading honeypot-based detection platform capable of detecting intrusions in real-time in ICSs/SCADA systems as well as IoT environments (Rubio et al. 2019). Its flagship product, BOTsink, is able to detect APTs without being detected by attackers. Other features also include software images to simulate SCADA devices as well as their protocols, so that they cannot be distinguished from real devices.

1.6.4.2.2. Industrial Defenica ICS Honeypot

In 2018, a worldwide study conducted by the Ponemon Institute on behalf of IBM found that the average time needed to identify a data breach is 197 days. Moreover, the most well-known attacks in industrial environments involved hackers who had penetrated inside the network for at least 3 months before being detected. Based on these results, Industrial Defenica proposed a high-interaction honeypot for ICS/SCADA industrial systems.

This solution employs advanced deception technology capable of presenting fake model-based units (PLC Ethernet-serial devices) in the network. More than 3,500 different devices can be tampered with (protocols, services, open ports, etc.), and these tampered devices can communicate with real ICS devices. Many steps are necessary for attackers to determine if they are dealing with a real device, and these steps would alert the security team that someone is breaking into the infrastructure.

To provide the best possible data about threats, a worldwide ICS Industrial Honeypot network was created to gather feedback from deployed solutions, as well as to improve data, equipment support, better equipment profiles so that they continue to appear as credible simulators.

1.6.4.3. Digital twins *for cyber-physical systems*

Digital twins were recognized as one of the top strategic technology trends of 2019 by Gartner (Eckhart and Ekelhart 2019). Here, we focus on a few use cases that demonstrate how they can harden the security of cyber-physical systems.

1.6.4.3.1. Definition

The standard definition of digital twins is that they are virtual replicas of physical objects that serve to monitor, visualize and predict the states of cyber-physical systems. Within the context of information security, Eckhart and Ekelhart (2019) has proposed a uniform definition based on the literature: a digital twin is "a virtual replica of a system that follows its physical counterpart during all the stages of its lifecycle, consumes real-time and history if necessary, and is realistic enough to allow the implementation of the desired security measure". Another term is also mentioned concerning digital twins: digital thread. In Eckhart and Ekelhart (2019), digital thread is defined as "the uninterrupted data link throughout a system lifecycle that can be used to generate and deliver updates to a digital twin".

1.6.4.3.2. Use cases in the manufacturing sector

Several use cases have been identified with digital twins operating to secure manufacturing systems:

- CPSs secure design;
- intrusion detection;
- misconfiguration detection (hardware and software);
- security tests;
- intimacy;
- system testing and training;
- secure dismantling;
- security and legal compliance.

To design more secure CPSs, digital twins should be used in combination with a virtual environment to analyze system behavior under attack. This would thus allow engineers to estimate potential damage, thereby facilitating the process of designing safety and security mechanisms to produce in the end more robust and fault-tolerant CPS architectures. Digital twins could also help identify weak points in the architecture or unnecessary features in devices that could contribute to exposing them to intrusion.

Digital twins can also help set up intrusion detection systems (IDSss). In this regard, Eckhart and Ekelhart (2019) presented a passive state replication approach that aimed at replicating a state from a physical device to a digital twin. This allows the digital twin to reflect the behavior of the actual CPS during its operation. Next, it is necessary to implement a behavior specification-based IDSs in which the normal CPS behavior is correctly defined to detect any changes. This technique presents low rates of false negatives and can detect certain attacks that were not known at the time the legitimate behavior was defined. Finally, intrusions can be simply detected by comparing the inputs and outputs of physical devices and their associated digital twins.

Since digital twins are the result of hardware and software emulation of devices, they mimic similar functionality. A hardware and software misconfiguration will be detected when different behaviors are observed between digital twins and their physical counterpart. If a difference is observed, it could indicate malicious actions. Unlike traditional configuration data analysis where only the software is verified, this use case also applies to hardware.

Security testing in operational environments (OT) is essential because it is performed on live systems and can cause serious damage or business interruption. Test benches are normally used to avoid interference, but their maintenance is costly in terms of time and effort. Digital twins enable security tests to be performed virtually instead of conducting them on real systems. How realistic a digital twin appears is a critical point, and this use case could also apply early in the engineering phase to fix vulnerabilities in the cyber-physical system. Using a digital twin as a honeypot system during the operational stages could also make it possible to test the security of the CPSs with real attackers and help harden the actual security of CPSss.

The concept of digital twins can also facilitate privacy protection, for example, by helping controllers or processors comply with the requirements of the General Data Protection Regulation (GDPR) (for instance, an insurer offering an insurance product based on data obtained from smart car digital twins). As digital twins use certain methods to classify data that can be anonymized before the data are transferred to the owner, privacy rights can be preserved.

Because digital twins are virtual and operate in an isolated environment, they can be used as a testing and training platform. The purpose of this platform could be to test new defenses or to train to control cyberattacks. The main idea is to launch attacks against digital twins from the virtual environment for testing and training purposes.

When the ICS and CPS reach end of life, the components must be disposed of in a secure manner. Several aspects have to be considered, such as data privacy requirements, as well as costs associated with disinfection. Digital twins could facilitate the secure disposal of physical devices. However, they can be affected by unauthorized access. It is therefore important to make sure that the digital thread is severed and correctly archived.

It seems that regulatory requirements for CPSs operators are increasing, which is the reason why digital twins could help by providing an in-depth reflection on CPSs throughout their lifecycle to enable continuous monitoring and documentation of security aspects.

1.7. Summary of good technical practices in cybersecurity

Security must also be addressed through appropriate technical capabilities and environments where they are deployed (ANSSI 2014; ENISA 2018).

This section will outline what we consider to be cybersecurity guidelines for Industry 4.0.

Table 1.7 summarizes these practices proposed by the ANSSI. For each practice, the following details are presented:

- the reason why it must be taken into account;
- the application method;
- its perimeter (hardware, networks, infrastructure, etc.);
- the application-related constraints;
- how to manage these constraints.

However, some IoT- and smart manufacturing-related metrics are missing. For this reason, they will be presented in detail in the following sections.

1.7.1. *Trust and integrity*

This practice makes it possible to guarantee data and device integrity and reliability. The following measures must be applied:

- verify software integrity and source before running it;
- for IoT devices, authorize them within the network using digital certificates/PKI;

Pratice	Motivation	Method	Perimeter	Constraints	Constraint management mechanisms
Physical access point control	Determine the system entry points	Identify access (Who? What for? How often?), protect servers, equipment, cable access (computer rooms, locked cabinets)	Workstations, servers, network devices, machines, touch screens	System size, maintain access even in an emergency	Alarmed door "dry contact", break glass procedures
Network segregation	Limit attack propagation and contain vulnerabilities	Establish a flow map, filter, trace and separate the network with VLANs	Networks (SCADA, PLC, development, etc.)	Real-time constraints (process networks)	Filtering applied upstream, physical access control to network equipment
Mobile device management	Reduce the risk of malware attacks with portable media (USB, HDD, etc.)	Define the policy, software restrictions, restrict use of this media, USB ports	Workstations, servers, consoles, touch screens	Need to exchange data between non-interconnected networks	Clean machines (strengthened, secured) for data transfers
Account management	Protect from unauthorized access	Account policy (users/applications), no default logins, strong passwords changed regularly	OS, databases, applications (SCADA/PLC), network devices, machinery	Generic accounts, emergency access	Traceability actions, strict procedures to determine identity with generic accounts (moment-by-moment)

Pratice	Motivation	Method	Perimeter	Constraints	Constraint management means
Configuration hardening	Restrict areas exposed to attacks	Install only necessary software, protocols and services, avoid default options, disable vulnerable protocols	OS, applications (SCADA/PLC), network devices, touch screens	Impact of alterations on production applications	Documented analysis, exception management for necessary non-secured features
Monitoring, warnings and alerts	Intrusion detection, action tracking, maintenance interventions	Enable traceability mechanisms (syslog, Windows events, etc.), filter and generate alerts for relevant events	OS, databases, network equipment, PLC, etc.	High volume of generated logs	Event management tools (filter, restrict, delete, etc.)
Configuration management	Ensure that there are no malicious modifications between versions	Comparison between running applications and reference application configuration, identify variations before deployment	Apps (SCADA/PLCs), network devices (configuration files)	ICSs complexity and heterogeneity	Configuration management tools to identify variations between two versions

Practice	Motivation	Method	Perimeter	Constraints	Constraint management mechanisms
Backup and restoring	Data must be available in case of full reboot after an attack or incident	Backup policy (which data should be backed up for users or according to regulatory requirements?)	Source codes, databases, logs, PLC firmware, network equipment configs (switches, routers, etc.)	Backup cannot be automatically performed for some devices (sensors, actuators for PLCs)	Track changes in settings, control, adjustments, sensor/actuator alarms
Documentation	Keep accurate system representations and control information dissemination	Documentation policy (update process, storage duration, etc.)	Factory, architectural plans, locations, administration and maintenance manuals, system analysis, etc.	Hard copies of documents may contain passwords (constraint), their control is complicated	Inform users of the risks with documentation, no document can be seen on the desktop
Malicious code detection	Advanced protection against virus attacks	Protection policy, priority to equipment/applications directly connected with the outside world and users	SCADA applications, engineering stations, programming and maintenance consoles	Incompatibility with older applications, no antivirus update, contractual issues (loss of warranty)	Deploy antivirus on handheld and maintenance machines, configuration hardening

Practice	Motivation	Method	Perimeter	Constraints	Constraint management mechanisms
Upgrade and patch management	Preventive protection against attacks, bug-related outages, vulnerabilities	Constraint-adapted patch management policy (systematic or periodic), identified risks and hardware	OS, applications, firmware, operator machines, servers, PLCs, telecom equipment, touch screens	Patches should be evaluated before deployment, some devices are not easily shut down	Identify vulnerabilities, schedule updates, monitor traffic, harden configurations and isolate devices
PLC protection	PLC program protection	PLC protected access (passwords), source code, read-only access for first-level maintenance, PLC cabinet locks	Production PLC, PLC programs	-	-
Engineering and development stations	Vulnerability points and contamination vectors (handhelds, connection to other networks)	Patches, antivirus software, no non-SCADA connection, usage monitoring, shutdown when not in use	SCADA development stations, PLC console, handheld devices for sensor and actuator configuration	-	-

Table 1.7. *Good cybersecurity practices for Industry 4.0*

– define secure data exchange channels for IoT devices (whitelist);

– implement application whitelists and periodic (annual) review of the list when changes occur;

– use cryptography mechanisms to ensure production data integrity;

– monitor data at rest and in transit to detect unauthorized modifications.

1.7.2. *Cloud security*

Security aspects of cloud computing are concerned by this practice. The following measures must be applied:

– choose the cloud type according to the company, the impact on privacy, the laws and the cloud provider's country;

– insert security and availability aspects in the agreements with cloud providers;

– avoid single point of failure with cloud applications and centralized systems;

– determine critical systems and applications when using public clouds;

– to minimize risk of cloud attacks, use a zero-knowledge approach and protect all data in the cloud and during transfer.

1.7.3. *Machine-to-machine safety*

The concept of machine-to-machine security is linked to key storage, encryption, input validation and protection during machine-to-machine communications. The following measures must be applied:

– use an HSM server in the infrastructure to store the service layer keys in the long term;

– establish a security association between communicating entities and cryptographic algorithms to ensure mutual authentication, integrity and confidentiality;

– use communication protocols capable of detecting unauthorized repetition of previous messages;

– for protection against cross-site scripting and command injection, use whitelist entry validation.

1.7.4. *Data protection*

Data protection is a very general concept, but in this case, it is related to the guarantee of data privacy at different levels within a company and managing access to data. The following measures must be applied:

– protection of data at rest (volatile and non-volatile memory), in transit and in use;

– data categorization based on risk analysis, criticality assessment and definition of security measures;

– grant access to specific data to third parties with least privileges and document this access;

– for highly confidential data, use encryption, key management and data loss prevention solutions;

– secure and anonymize direct and indirect personal data processed through access controls, roles and encryption.

1.7.5. *Software/firmware updates*

Just as with any other device, software updates with IoT solutions must follow a specific methodology, whose measures are:

– ensure strict update control (no modifications between the source and the destination);

– only deploy patches after testing and proving that there are no negative consequences;

– for systems that cannot be updated, compensatory measures must be applied.

1.7.6. *Access control*

IoT devices are critical devices in Industry 4.0 and can be physically or remotely accessed. For this reason, the following access control measures must be applied:

– minimum authentication and authorization level for given system segments;

– implementation of multi-factor authentication;

– application of the principle of least privilege;

– implementation of an account lockout feature;

– remote access segregation;

– physical access security must be taken into consideration.

1.7.7. *Networks, protocols and encryption*

Proper protocol implementation, encryption and network segmentation are key elements for building a strong IoT network. The following actions are linked to this objective:

– use secure communication and encryption channels when possible;

– use proven standards-based protocols (TLS 1.3) and avoid vulnerable protocols (Telnet, SNMP v1/v2);

– limit the number of protocols in a given environment and disable unused services;

– ensure security capabilities and interoperability between protocols.

1.7.8. *Monitoring and auditing*

Several measures are suggested for the IoT environment regarding network traffic, availability monitoring and log analysis:

– implementation of a passive monitoring solution to create a baseline of industrial network traffic;

– real-time security log analysis using SIEM solutions (Security Information and Event Management);

– periodic log, access privilege and configuration reviews;

– real-time availability monitoring of IoT devices.

1.7.9. *Configuration management*

The IoT encompasses a wide range of devices; it is important to keep track of changes in configurations, device hardening and backup methods.

The following measures are in line with this goal:

– basic customized security configurations for different types of assets;

– documentation of any configuration changes in accordance with the change management policy of the organization;

– implementation of support tools for configuration management;

– creation of a complete backup plan, adapted to the different types of assets.

1.8. Industry 4.0, cybersecurity and traceability outlines

In this chapter, we have explored cybersecurity concepts and solutions within the context of Industry 4.0 through scientific and normative literature. In the analyzed articles, certain aspects of cybersecurity were not always mentioned, such as the managerial aspect of cybersecurity or the business impacts (Zia et al. 2022). For this reason, our study was structured as a step-by-step approach to provide a comprehensive overview of the topic. In addition to being an introductory review of scientific work on cybersecurity, it can also be used as a guide to introducing cybersecurity into a smart manufacturing environment.

After introducing the concept of Industry 4.0, we showed the complexity created by cybersecurity mechanisms to ensure system security, due to the many innovative technologies involved.

We then proposed a characterization of cybersecurity in terms of both its technical and managerial aspects to show that every level in an organization has a role to play. Given that cybersecurity must be considered in terms of physical zones, we proposed a division of a factory into several perimeters, which we characterized according to their interactions, equipment and networks. We then outlined the cybersecurity vulnerabilities, threats and risks met in Industry 4.0 factories for each perimeter, while taking into account the business impacts at the organizational level (Figure 1.12).

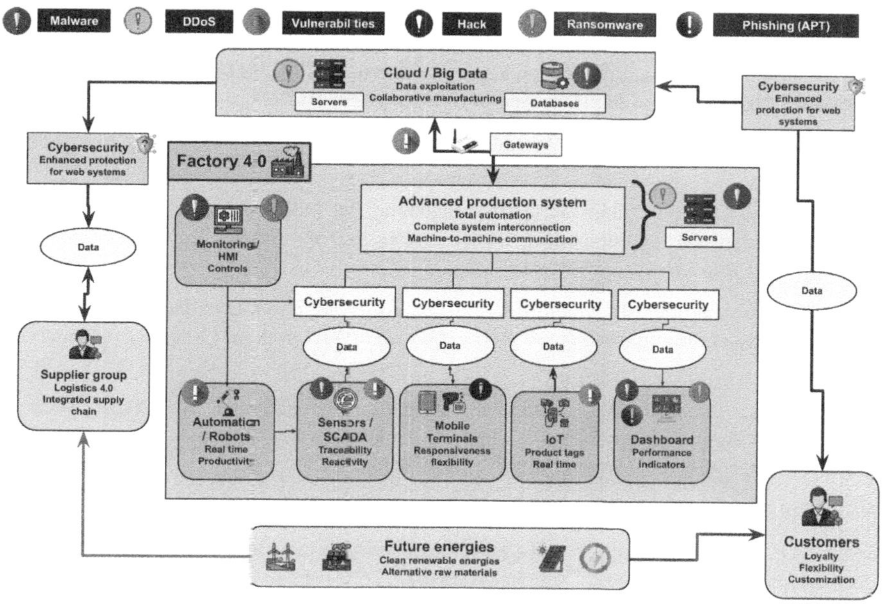

Figure 1.12. *Cybersecurity risks overview in a factory 4.0*

On the one hand, connecting the factory to the Internet in order to enable real-time information transmission and sharing with the outside world will result in the creation of a gateway between networks, which may give rise to new security breaches. On the other hand, the factory devices each exhibit their own risks in terms of cybersecurity (mobile devices can be hacked, the IoT has vulnerabilities, servers are prone to DDoS, etc.). Among these new technologies, some are closely related to the notions of product identification and traceability in the factory (sensors, IoT, RFID tags, etc.).

It can be seen that Factory 4.0 will be managed by data from different sources, both internal and external. With the help of outsourced storage through big data and the Cloud, the factory will be able to manage all the data related to a product, from its design to its end of life, from its production to its marketing, and be able to keep them for as long as necessary. Most importantly, the factory will need to be able to retrieve, integrate and leverage data created by the industry in general and its customers in particular. Once all these data have been compiled, it will be able to infer a consistent predictive analysis. Chapter 2 will address the topic of the operational management of the part of this mass of data created by the traceability system.

1.9. References

Abu Jassar, A., Attar, H., Yevsieiev, V., Amer, A., Demska, N., Luhach, A., Lyashenko, V. (2022). Electronic user authentication key for access to HMI/SCADA via unsecured internet networks. *Computational Intelligence and Neuroscience*, 2022(1), 1–13.

Alani, M. and Alloghani, M. (2019). *Security Challenges in the Industry 4.0 Era*. Springer, Berlin.

Alcácer, V. and Cruz-Machado, V. (2019). Scanning the industry 4.0: A literature review on technologies for manufacturing systems. *Engineering Science and Technology, an International Journal*, 22(3), 899–919.

Ali, A., Shah, G., Farooq, M., Khan, M.U. (2017). Technologies and challenges in developing machine-to-machine applications: A survey. *Journal of Network and Computer Applications*, 83, 1–13.

Alqudhaibi, A., Albarrak, M., Jagtap, S., Williams, N., Salonitis, K. (2025). Securing Industry 4.0: Assessing cybersecurity challenges and proposing strategies for manufacturing management. *Cyber Security and Applications*, 3, 100067.

ANSSI (2014). Managing cybersecurity for industrial control systems. Report, ANSSI.

Asghar, M.R., Hu, Q., Zeadally, S. (2019). Cybersecurity in industrial control systems: Issues, technologies, and challenges. *Computer Networks*, 165, 106946.

Ashton, K. (2009). That 'Internet of Things' thing. *RFiD Journal*, 22, 97–114.

Atzori, L., Iera, A., Morabito, G. (2016). Understanding the Internet of Things: Definition, potentials, and societal role of a fast evolving paradigm. *Ad Hoc Networks*, 56, 122–140.

Bhamare, D., Zolanvari, M., Erbad, A., Jain, R., Khan, K., Meskin, N. (2020). Cybersecurity for industrial control systems: A survey. *Computers & Security*, 89, 101677.

Brandao, P. and Duarte, P. (2022). Cybersecurity risk management in the Industry 4.0. *International Journal of Scientific Research And Growth*, 9, 661–668.

Carcano, A., Coletta, A., Guglielmi, M., Masera, M., Nai Fovino, I., Trombetta, A. (2011). A multidimensional critical state analysis for detecting intrusions in scada systems. *IEEE Transactions on Industrial Informatics*, 7(2), 179–186.

Corallo, A., Lazoi, M., Lezzi, M. (2020). Cybersecurity in the context of industry 4.0: A structured classification of critical assets and business impacts. *Computers in Industry*, 114, 103165.

Dawson, M. (2018). Cyber security in Industry 4.0: The pitfalls of having hyperconnected systems. *Journal of Strategic Management Studies*, 10(1), 19–28.

Eckhart, M. and Ekelhart, A. (2019). *Digital Twins for Cyber-Physical Systems Security: State of the Art and Outlook*. Springer, Berlin.

ENISA (2018). Good practices for security of Internet of Things in the context of smart manufacturing. Report, ENISA.

European Commission (2021). Industry 5.0 – Towards a sustainable, human-centric and resilient European industry [Online]. Available at: https://research-and-innovation.ec.europa.eu/knowledge-publications-tools-and-data/publications/all-publications/industry-50-towards-sustainable-human-centric-and-resilient-european-industry_en.

Fernández-Caramés, T.M., Fraga-Lamas, P. (2019). A review on the application of blockchain to the next generation of cybersecure Industry 4.0 smart factories. *IEEE Access*, 7, 45201–45218.

Flaus, J.-M. (2019). *Cybersecurity of Industrial Systems*. ISTE Ltd, London and Wiley, Hoboken, NJ.

Geandarme, F. (2018). Industrie 4.0 : définition et mise en œuvre vers l'usine de production connectée [Online]. Available at: https://www.visiativ.com/actualites/actualites/industrie-4-0-definition-et-mise-en-oeuvre-vers-lusine-de-production-connectee/.

Graham, J., Hieb, J., Naber, J. (2016). Improving cybersecurity for industrial control systems. In *Proceedings of the 2016 IEEE 25th International Symposium on Industrial Electronics (ISIE)*, Santa Clara, CA.

Haleem, A., Javaid, M., Singh, R., Rab, S., Suman, R. (2022). Perspectives of cybersecurity for ameliorative Industry 4.0 era: A review based framework. *Industrial Robot*, 49(3), 582–597.

Hassija, V., Chamola, V., Saxena, V., Jain, D., Goyal, P., Sikdar, B. (2019). A survey on IoT security: Application areas, security threats, and solution architectures. *IEEE Access*, 7, 82721–82743.

Illési, Z., Halász, A., Varga, P.J. (2018). Wireless networks and critical information infrastructure. In *Proceedings of the 2018 IEEE 12th International Symposium on Applied Computational Intelligence and Informatics (SACI)*. Forschungsunion, Acatech, Munich.

Kagermann, H., Wahlster, W., Helbig, J. (2013). Recommendations for implementing the strategic initiative Industrie 4.0 [Online]. Available at: https://www.din.de/resource/blob/76902/e8cac883f42bf28536e7e8165993f1fd/recommendations-for-implementing-industry-4-0-data.pdf.

Kiss, M., Breda, G., Muha, L. (2019). Information security aspects of Industry 4.0. *Procedia Manufacturing*, 32, 848–855.

Kondiloglu, A., Bayer, H., Celik, E., Atalay, M. (2017). Information security breaches and precautions on Industry 4.0. *Tehnologčnij Audit ta Rezervi Virobnictva*, 64(38), 58–63.

Lezzi, M., Lazoi, M., Corallo, A. (2018). Cybersecurity for Industry 4.0 in the current literature: A reference framework. *Computers in Industry*, 103, 97–110.

Lin, J., Yu, W., Zhang, N., Yang, X., Zhang, H., Zhao, W. (2017). A survey on Internet of Things: Architecture, enabling technologies, security and privacy, and applications. *IEEE Internet of Things Journal*, 4(5), 1125–1142.

Liu, H. and Lang, B. (2019). Machine learning and deep learning methods for intrusion detection systems: A survey. *Applied Sciences*, 9, 4396.

Lu, T., Guo, X., Li, Y., Peng, Y., Zhang, X., Xie, F., Gao, Y. (2014). Cyberphysical security for industrial control systems based on wireless sensor networks. *International Journal of Distributed Sensor Networks*, 10(6), 438350.

McLaughlin, S., Konstantinou, C., Wang, X., Davi, L., Sadeghi, A., Maniatakos, M., Karri, R. (2016). The cybersecurity landscape in industrial control systems. *Proceedings of the IEEE*, 104(5), 1039–1057.

Miraz, D., Ali, M., Excell, P., Picking, R. (2015). A review on Internet of Things (IoT), Internet of Everything (IoE) and Internet of Nano Things (IoNT). In *Proceedings of 2015 Internet Technologies and Applications (ITA)*, Wrexham, 219–224.

Mitchell, R. and Chen, I. (2013). Behavior-rule based intrusion detection systems for safety critical smart grid applications. *IEEE Transactions on Smart Grid*, 4(3), 1254–1263.

Mozzaquatro, B.A., Jardim-Goncalves, R., Agostinho, C. (2015). Towards a reference ontology for security in the Internet of Things. In *Proceedings of the IEEE International Workshop on Measurements Networking*, Coimbra, 1–6.

Mozzaquatro, B.A., Agostinho, C., Goncalves, D., Martins, J., Jardim-Goncalves, R. (2018). An ontology-based cybersecurity framework for the Internet of Things. *Sensors*, 18(9), 3053.

Mullet, V., Sondi, P., Ramat, E. (2021). A review of cybersecurity guidelines for manufacturing factories in industry 4.0. *IEEE Access*, 9, 23235–23263.

Nunes, B.A.A., Mendonca, M., Nguyen, X., Obraczka, K., Turletti, T. (2014). A survey of software-defined networking: Past, present, and future of programmable networks. *IEEE Communications Surveys Tutorials*, 16(3), 1617–1634.

Omari, A., Alsariera, Y., Alhadawi, H., Albawaleez, M., Alkhliwi, S. (2022). A closer look on challenges and security risks of voice over internet protocol infrastructures. *IJCSNS*, 22, 175.

Ozarpa, C., Avci, I., Eldekçi, A. (2022). Industry 4.0 and cybersecurity at automobile manufacturing in smart factories [Online]. Available at: https://www.researchgate.net/publication/360836993_Industry_40_and_Cybersecurity_at_Automobile_Manufacturing_in_Smart_Factories_Akilli_Fabrikalardaki_Otomobil_Imalatinda_Endustri_40_ve_Siber_Guvenlik.

Pan, Y., White, J., Schmidt, D., Elhabashy, A., Sturm, L., Camelio, J., Williams, C. (2017). Taxonomies for reasoning about cyber-physical attacks in IoT-based manufacturing systems. *International Journal of Interactive Multimedia and Artificial Intelligence*, 4, 45–54.

Pereira, T., Barreto, L., Amaral, A. (2017). Network and information security challenges within Industry 4.0 paradigm. *Procedia Manufacturing*, 13, 1253–1260.

Petrenko, S. (2020). *Developing a Cybersecurity Immune System for Industry 4.0*. River Publishers, London.

Prinsloo, J., Sinha, S., von Solms, B. (2019). A review of Industry 4.0 manufacturing process security risks. *Applied Sciences*, 9(23), 5105.

Rahman, Z., Yi, X., Khalil, I. (2022). Blockchain based ai-enabled Industry 4.0 CPS protection against advanced persistent threat. *CoRR*. doi: 10.48550/arXiv.2201.12727.

Reinsel, D., Gantz, J., Rydning, J. (2018). The digitization of the world from edge to core. Report, Seagate.

Rubio, J.E., Alcaraz, C., Roman, R., Lopez, J. (2019). Current cyber-defense trends in industrial control systems. *Computers and Security*, 87, 101561.

Rudenko, R., Pires, I., Oliveira, P., Barroso, J., Reis, A. (2022). A brief review on Internet of Things, Industry 4.0 and cybersecurity. *Electronics*, 11, 1742.

Satasiya, D. and Raviya Rupal, D. (2016). Analysis of software defined network firewall (SDF). In *Proceedings of the 2016 International Conference on Wireless Communications, Signal Processing and Networking (WiSPNET)*. IEEE, Chennai, 228–231.

Sawik, T. (2020). A linear model for optimal cybersecurity investment in Industry 4.0 supply chains. *International Journal of Production Research*, 60(4), 1368–1385.

Shukla, M., Sarmah, S., Tiwari, M. (2022). A multi-objective framework for the identification and optimisation of factors affecting cybersecurity in the Industry 4.0 supply chain. *International Journal of Production Research*, 61(15), 5266–5281.

Sicard, F., Zamai, É., Flaus J.-M. (2018). *Critical States Distance Filter Based Approach for Detection and Blockage of Cyberattacks in Industrial Control Systems*. Springer, Berlin.

Sicard, F., Zamai, E., Flaus, J.-M. (2019). An approach based on behavioral models and critical states distance notion for improving cybersecurity of industrial control systems. *Reliability Engineering and System Safety*, 188, 584–603.

Tange, K., De Donno, M., Fafoutis, X., Dragoni, N. (2020). A systematic survey of industrial Internet of Things security: Requirements and fog computing opportunities. *IEEE Communications Surveys Tutorials*, 22(4), 2489–2520.

Thames, L. and Schaefer, D. (2017). *Cybersecurity for Industry 4.0 and Advanced Manufacturing Environments with Ensemble Intelligence*. Springer, Berlin.

Tsuchiya, A., Fraile, F., Koshijima, I., Ortiz, A., Poler, R. (2018). Software defined networking firewall for Industry 4.0 manufacturing systems. *Journal of Industrial Engineering and Management*, 11(2), 318–333.

Wegner, A., Graham, J., Ribble, E. (2017). *A New Approach to Cyberphysical Security in Industry 4.0*. Springer, Berlin.

Whitmore, A., Agarwal, A. Xu, L. (2014). The Internet of Things – A survey of topics and trends. *Information Systems Frontiers*, 17, 261–274.

Wolf, M. and Serpanos, D. (2018) Safety and security in cyber-physical systems and Internet-of-Things systems. *Proceedings of the IEEE*, 106(1), 9–20.

Zeng, P., Wang, Z., Jia, Z., Kong, L., Li, D., Jin, X. (2019). Time-slotted software-defined industrial ethernet for real-time quality of service in Industry 4.0. *Future Generation Computer Systems*, 99, 1–10.

Zhao, G., Zhang, C., Zheng, L. (2017). Intrusion detection using deep belief network and probabilistic neural network. In *Proceedings of the 2017 IEEE International Conference on Computational Science and Engineering (CSE) and IEEE International Conference on Embedded and Ubiquitous Computing (EUC)*, 639–642.

Zia, N., Burita, L., Huseynova, A., Owusu, V. (2022). A managerial review and guidelines for Industry 4.0 factories on cybersecurity. *European Conference on Cyber Warfare and Security*, 21, 336–340.

2

Global Approach to Traceability for Factory 4.0

Traceability is a broad concept that refers to the practice of identifying an object or working component to access part or all of its related information anywhere during its lifecycle. With the advent of Industry 4.0 and major changes in the way industrial processes operate, the concept of traceability involves a strategic notion that goes further than merely identifying products and their location within the factory. Technologies such as the cloud or the IoT enable exchanges to be achieved between the factory traceability system and partners; these include customers and suppliers in order to foster trust between them by guaranteeing product quality and safety.

However, the resulting transparency cannot be introduced at the expense of undesirable exposure of confidential factory data. This chapter will therefore present and outline our approach to product-oriented traceability and how trust can be improved between partners without jeopardizing data privacy.

The content of sections 2.1 to 2.3 covers established knowledge on the field and is virtually the same from one source to another. In this book, it draws heavily from the research proposed in Keyence (2024) and Scantrust (2024).

2.1. Basic principles

The growing importance that quality and security globally acquired has contributed to the fact that traceability should now be considered of the utmost importance and has gained increased attention from many industrial sectors, including automotive, electronic components, food and pharmaceuticals.

For a color version of all the figures in this chapter, see: www.iste.co.uk/sondimullet/cybersecurity.zip

Cybersecurity and Traceability in Factory 4.0,
coordinated by Patrick SONDI and Valentin MULLET.

The different definitions of the notion of traceability can generally be summarized into two aspects: chain traceability and internal traceability.

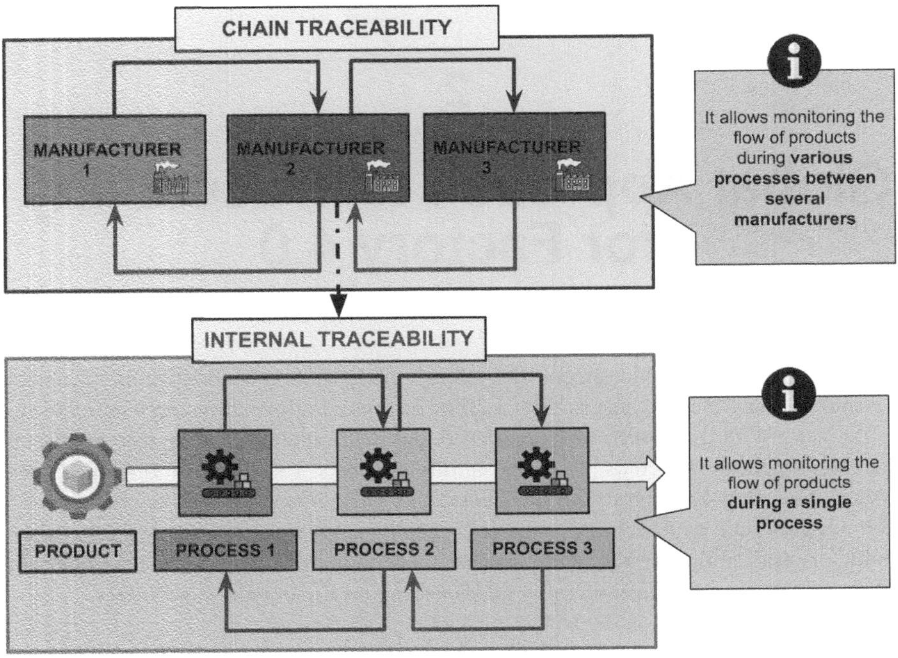

Figure 2.1. *Illustration of chain traceability and internal traceability*

2.1.1. *Chain traceability*

The general understanding of traceability at the global level corresponds to chain traceability. The latter embodies the possibility of upstream or downstream tracking of the history from the supply of raw materials and parts to the sale, including machining, assembly and distribution (Figure 2.2).

Manufacturers can track the location where products are delivered (downstream tracking), while companies and consumers located downstream can visualize where the products they access are coming from. Manufacturers can thus benefit from a simpler identification of root causes of faults allowing them to more easily recall their products in the event of unforeseen problems. Thereby, consumers can also monitor and seemlessly select reliable products without labeling errors.

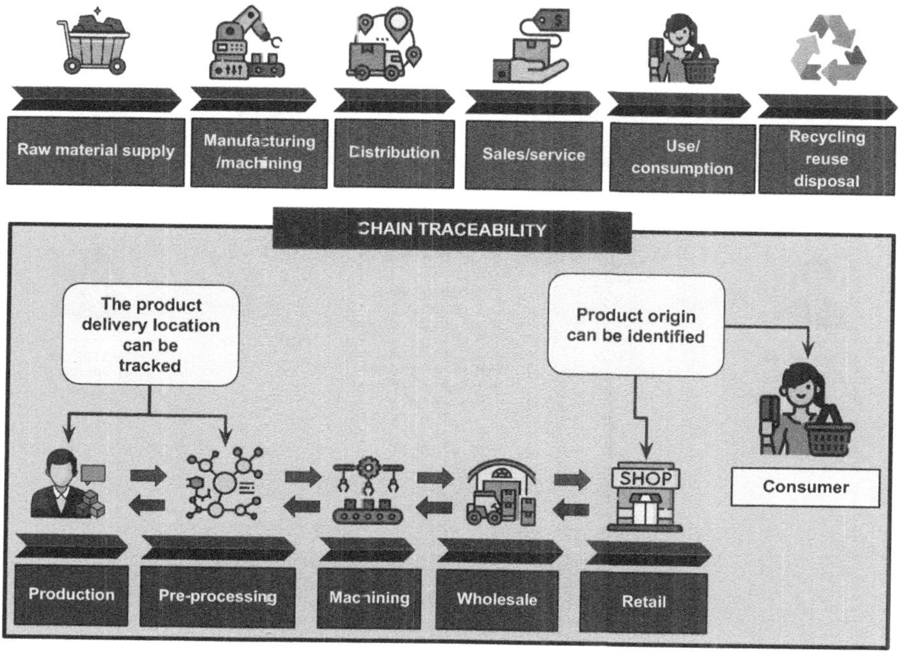

Figure 2.2. *Chain traceability overview*

2.1.2. *Internal traceability*

Internal traceability refers to tracking the product journey within an identified perimeter as part of a global supply chain, such as a factory.

For example, an engine assembly plant sources engine parts such as camshafts and pistons from its suppliers and assembles them. The management and use of the manufacturing history as well as the inspection results of these parts by the factory can also be designated as falling under the scope of internal traceability.

2.1.3. *Examples of internal traceability*

2.1.3.1. *Traceability of manufacturing processes*

Traceability of manufacturing processes includes collecting and processing information related to the operations carried out during the manufacturing processes, from the moment raw materials and components are delivered to the shipping of manufactured products.

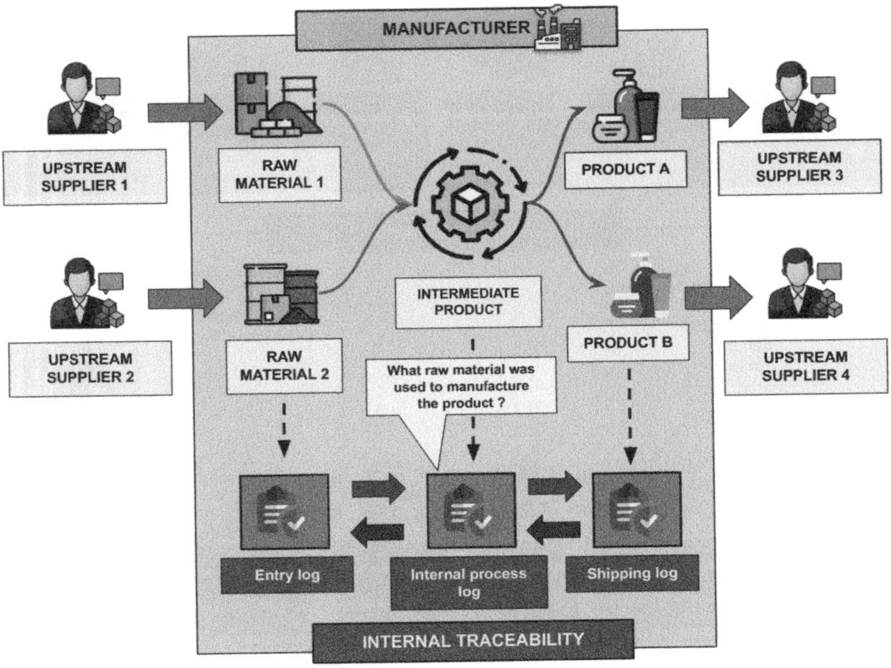

Figure 2.3. *Internal traceability overview*

Traceability of manufacturing processes assigns each product or batch an identification number. For each process, information such as task details, inspection outcomes and dimensions are attached to this identification number so that subsequent processes can consult it if needed (Figure 2.4). In general, this information can be used during machining to help improve efficiency and quality.

2.1.3.2. *Part inspection traceability*

Part management traceability simplifies the control and analysis of recurring wear and tear of parts such as tools and jigs. A serial number is defined for individual identification (such as a 2D code) and inscribed on each tool in order to manage its condition, its service life and wear limits in particular. As with the entire flow, tools and other stored parts are marked with 2D management codes and information is assigned to them, namely the factory name, shelf number and serial number, for the management of releases and returns. Further information, such as the number of grinding operations in addition to their manufacturing date and time, is collected and managed to ensure and standardize product quality.

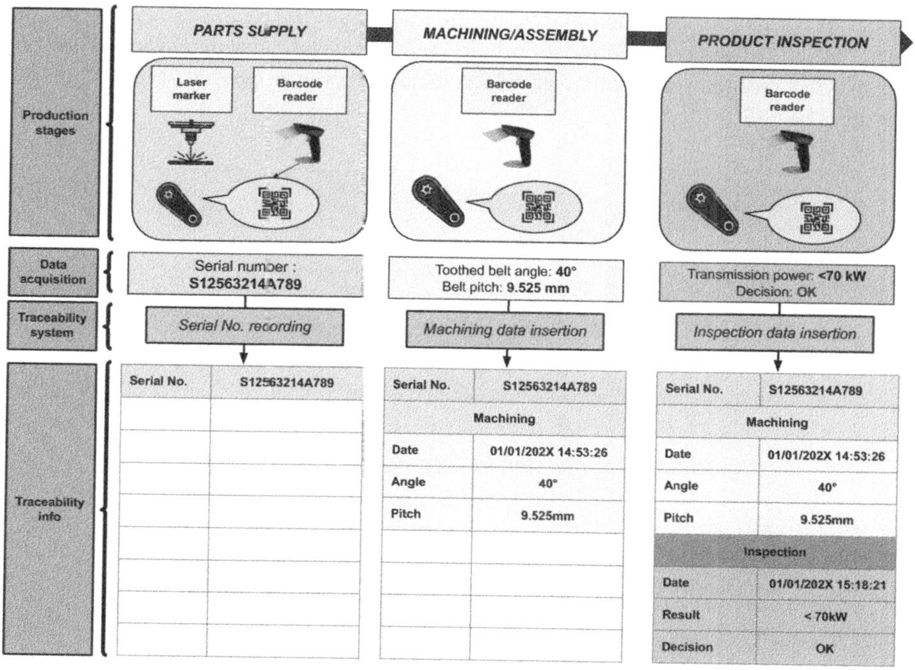

Figure 2.4. *Internal traceability during several manufacturing processes*

2.1.4. *Downstream and upstream traceability*

Products are identified either individually or in batches when collecting information for each process. Downstream tracing means that accumulated information is used to track the product journey and upstream tracing designates the monitoring of records by tracking back the product timeline. Nonetheless, more should be done besides identifying parts, products and accumulating information. Traceability is only ensured if this information is accessible and can be traced downstream and upstream at any time.

2.1.4.1. *Downstream traceability*

Downstream traceability consists of tracing a product by following the timeline of its production. For example, when a fault is detected in certain parts, the products containing these parts can be identified to carry out a targeted recall. Traceability is therefore an effective measure for preventing recalls of faulty products (Figure 2.5).

2.1.4.2. *Upstream traceability*

Upstream traceability consists of following a product by tracking back in time. When a problem is detected involving received products, the batch and the process concerned should be identified by tracing the manufacturing record to investigate the cause. Batch or process identification helps to quickly take the necessary measures to solve the problems and thus ensure a constant or even improved level of quality (Figure 2.6).

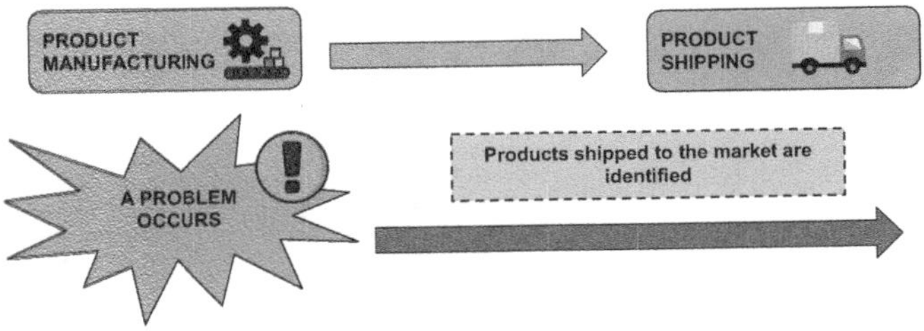

Figure 2.5. *Downstream traceability visual representation*

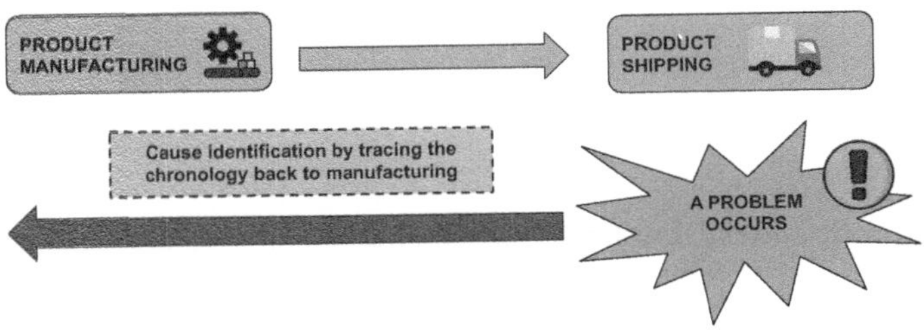

Figure 2.6. *Upstream traceability visual representation*

2.1.5. *Identification formats*

The implementation of upstream and downstream traceability is based on collecting and recording appropriate information at each process level.

The required information depends on several factors, including the type of industry, products and processes. In the manufacturing industry, it is important to know the sender, the manufacturing details, the manufacturing and inspection dates and times, the manager, the production line and the recipient. In the food industry, the producer, the place of production, the dates and times of shipment and processing, the expiration date and finally the recipient should be the information to consider. Table 2.1 shows a list of traceability information sorted by information source.

Information source	Examples of traceability information
Delivery	Delivered quantity Delivery date and time Serial/lot no. Producer, manufacturer, etc., information
Manufacturing/machining	Manufactured/processed quantity Manufacturing/processing date and time Serial/lot no. Manufacturing history Inspection history
Shipping	Shipped quantity Shipping date and time Serial/lot no. Customer information Production factory information

Table 2.1. *Examples of information to be collected to establish traceability*

2.1.5.1. *Definition of identification symbols*

In order to ensure efficient information processing, symbols and rules should be established for the definition of identification formats. When the established rules are not followed and no distinction is made between the symbols used by these formats, the information cannot be accurately identified, thus compromising upstream and downstream traceability.

When several suppliers or manufacturers are involved in the production of a product, it becomes essential to establish a definition rule for the symbols and identification formats common to all the companies concerned.

2.1.5.2. *Representation formats and transfer media*

There is a wide range of identification symbols employed to transfer information. It ranges from numbers and characters easily understood by humans to barcodes, 2D codes and electronic signals. These symbols and the associated combination methods are called "representation formats". They only refer to symbols such as characters or 2D codes. They are associated with a transfer media (storage media) such as a paper label or a smart tag attached to the product or part.

For example, suppose that we want to stick a label on a box containing a delivered part. The delivery date and the name of the product are inserted with a pen on the box. In this example, the label is the transfer medium and the characters indicating the delivery date and the product name are the representation format. If this information is directly written on the cardboard, the transfer medium is the box itself. On the other hand, if the data are directly written on the part using a laser, the characters are the representation format and the transfer medium is transparent, because it is the product itself.

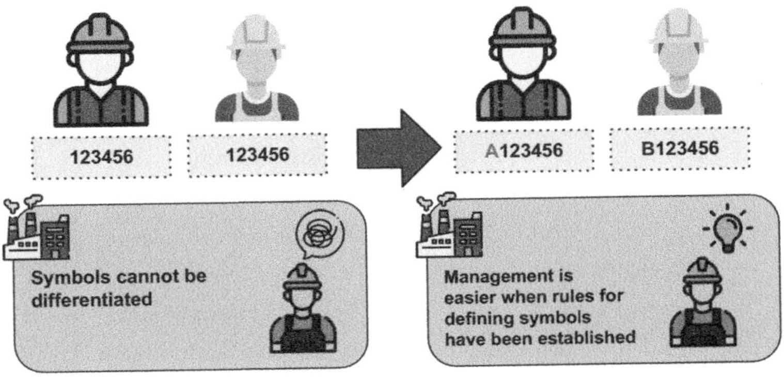

Figure 2.7. *Definition of identification formats*

2.1.6. *Data collection*

Information can be attached to a part or product by means of a representation format and a transfer medium. However, to ensure traceability, not only should this information be attached, it must be recorded and managed.

2.1.6.1. *Recording media*

As a result, it is necessary to read the representation format on the transfer medium, record the information it contains and manage it. Paper registers, computers or cloud servers are all media for recording and managing information. Within the traceability framework, they are identified as recording media.

Information can be collected and managed either by copying it by hand into a ledger or by typing it or by scanning it and transferring it with a code reader onto a server. Once the rules allowing information retrieval when necessary have been established, upstream and downstream traceability can be implemented.

2.1.6.2. *Traceability data collection use cases*

2.1.6.2.1. Character string and hard copy

The system uses characters (letters and numbers) that can be recognized by the human eye as representation format, while labels or stickers are the transfer means.

For example, numbers such as *20240101* to communicate the expiration date, a character string such as *LOT000101* to indicate a batch number or *ABZ0101* to specify the product serial number. Paper hard copies used as transfer media can be divided into two categories: media inserted on the product, such as labels and packaging, and media included with the product, such as invoices and delivery notes, for instance. Concerning the storage medium, paper-based records are usually employed. However, direct storage on a computer using a hand-held terminal with Optical Character Recognition (OCR) functionality is becoming increasingly popular.

2.1.6.2.2. Barcode and paper-based

The system uses barcodes as a representation format and they are printed on labels or packaging that serve as transfer media. As barcodes cannot be understood by the human eye, a code reader or other scanning device is required. A printer and software are also required to print the barcode onto a label or other transfer media. The storage medium is a computer, which allows for simple information management, contactless reading and implementation with a relatively low cost. This combination is used in a variety of industries, and in the manufacturing, food, pharmaceutical and distribution sectors in particular.

2.1.6.2.3. 2D code and direct marking

The use of 2D codes as representation format has become widespread in recent years. With 2D codes, a larger amount of information can be inserted in a smaller area. In addition, the reading accuracy is constantly being improved. Direct marking consists of directly inserting the code on the part. The fact that no transfer medium is used guarantees reduced operational costs and decreases dissociation/substitution risks.

2.1.6.2.4. Electronic data and smart tag (RFID Tag)

Electronic data are used as representation format and small tags with a microchip (RFID tag) as transfer media. Smart tags communicate information via radio waves. Their main advantage is that information acquisition is even possible remotely or on several labels simultaneously. However, their main disadvantage remains their high cost.

Table 2.2 summarizes the different collection methods presented above, highlighting their characteristics as well as their advantages and disadvantages.

2.1.7. *Traceability challenges and significance*

If a product has quality defects, the manufacturer of that product must take effective and prompt action. A slow or ineffective response from this manufacturer can create mistrust among consumers or business partners, which can strongly impact the company. Consumer protection legislation has evolved in such a manner that an increasing number of companies have been driven to quickly recall their products due to problems.

Use cases	Representation format	Transfer media	Recording media	Advantages +	Disadvantages -
Barcode and paper-based			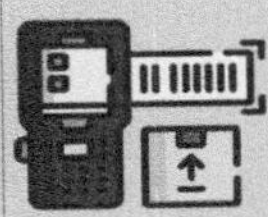	Simple information management Contactless reading Low cost	- Durability/longevity - Can only be read with a reader - Limited storage capacity
2D code and direct marking				High storage capacity Reading accuracy No transfer media (cost reduction)	- Depending on the marking method (non-persistent, usable area deterioration or reduction, etc.)
Electronic data and smart tag (RFID tag)		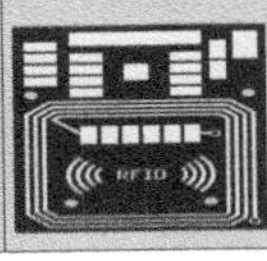		Remote data acquisition Simultaneous reading of several labels	- High cost

Table 2.2. *Summary of traceability data collection use cases*

Strongly recommended in the automotive industry for a long time, traceability is widely used for avoiding products recalls, minimizing damage, defect identification/solution and quality management purposes. On the other hand, tracking data from tens of thousands of components from manufacturing to disposal, while complying with constantly changing laws and regulations is a difficult task. Traceability is nonetheless a major issue in globalized trade, governed by fierce competition in terms of price, time and service quality. Virtually all industry sectors now feel that they are compulsorily required to implement an exhaustive history management system, which would cover both internal and external traceability.

2.2. Implementation principles

This section describes in a structured way the procedures for implementing traceability based on a typical example of production flow in the manufacturing industry. The first step consists of identifying parts and products, and then managing them by attaching objects and information to them. Key concepts are therefore identification and link management. A suitable description therefore begins with the method for the formalization of object flow, before addressing identification.

2.2.1. *Object flow formalization*

Traceability can be understood from two angles, namely *internal traceability* and *chain traceability*. The example of internal traceability under consideration is viewed at the limited structural scale such as a factory, which is the main subject of this book.

When implementing traceability, the production process flow should be described by satisfying the following points:

1) Are raw materials and parts supplied by a single entity or more than one third party?

2) Who verifies that the delivery order matches the received materials and parts and how?

3) What is the stock management method (individual or batch-based)?

4) How are machining, assembly and inspection reports registered and managed?

5) What information is required and what information should be captured?

2.2.2. *Identification*

Identification essentially consists of recognizing raw materials, parts and products individually or in batches. Its purpose is to enable recognition of each product, even when similar products with the same shape and material are mass produced. This can be achieved in two different manners, either by assigning individual identification symbols to each part or product, or by defining a specific group as a batch and assigning an identification symbol to the batch. Individual items or groups that have been assigned an identification symbol are called identification units.

2.2.2.1. *Individual identification*

Each part or product is assigned a unique serial number and is individually managed (Figure 2.8). When recall is required, faulty products can be reliably located. This method is also called serial number-based management because it makes use of serial numbers as unique identifiers.

2.2.2.2. *Batch identification*

Products manufactured under identical conditions are considered to belong to the same group called a *batch* and each batch is assigned an identification number. This management method requires verification that the quality of all products is uniform. Since all the products in the batch share the same identification number, it is impossible to individually identify them (Figure 2.9).

Batch-based management is a means of implementing traceability at a lower cost because it increases the identification range. Conversely, for increased accuracy, smaller batches should preferably be built or individual management be considered.

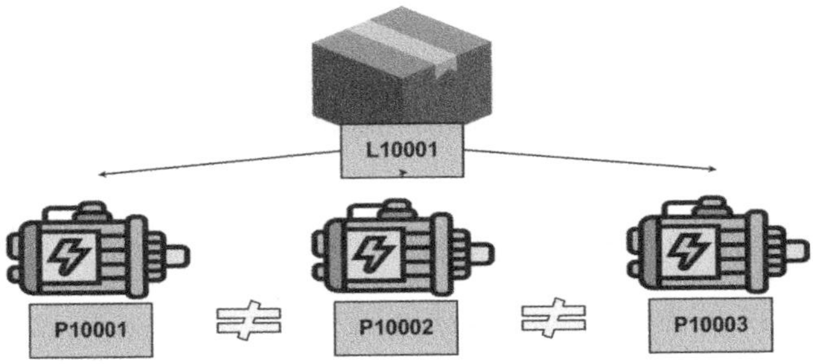

Figure 2.8. *Graphical representation of individual identification*

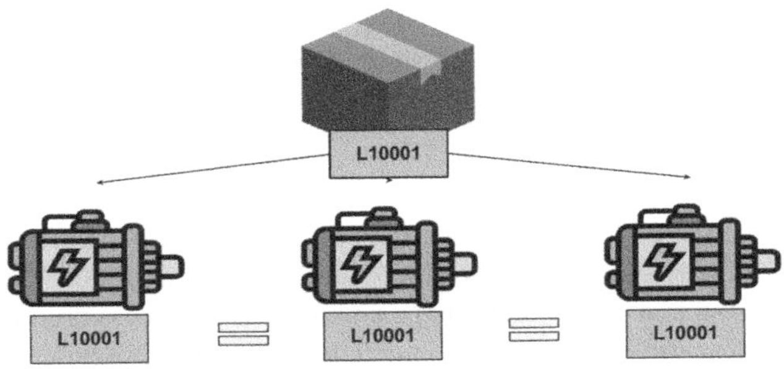

Figure 2.9. *Graphical representation of batch identification*

In batch management, the tracking accuracy varies according to the criteria for distinguishing batches. The easiest way to create a batch consists of grouping identical products made on the same day, without changing the manufacturing parameters. For example, if 1,000 parts are manufactured in one day, they should be grouped in a single batch and the same batch number should be used for each one. Parts or products from the same batch must be manufactured with the same machine, on the same day and with the same quality.

Any changes in conditions during production is likely to affect product quality, whether they concern machine settings, the manufacturer supplying the parts, or the person in charge of manufacturing. In this case, the batch number (batch reduction) should also be changed. The smaller a batch, the more limited the number of defective products in a potential recall. This management method by batch division is called segregation management. Nonetheless, due to segregation and higher costs, the task is then more difficult. It therefore becomes essential to study cost-effectiveness when creating batches.

2.2.3. *Link/association management*

Link management consists of associating one object to another, an object to information or an information element to another. Suppose that a factory manufactures an engine in which parts of different sizes, such as crankshafts and pistons, are assembled. Link management refers to the set of rules governing the accumulation, organization and retrieval of the established information used for creating records, namely the location where the piston was manufactured or the machining process used for the connecting rod from the identification symbol. Also called association, it is essential for the implementation of downstream and upstream traceability.

2.2.3.1. *Ensuring one-tier upstream or downstream traceability*

The detailed presentation of link management will be based on an example of production processes. We assume that traceability up to parts suppliers (one tier upstream) and down to customers (one tier downstream) is ensured.

In order to ensure upstream traceability, the following steps must be followed:

– The date, sender and quantity should be verified on the suppliers' delivery notes for each received part.

– Identification numbers have to be assigned to the parts linked to this information.

– Information data should be registered and stored within the factory.

When a part is defective, this method makes it possible to follow the recorded information back to the supplier.

Regarding customer and downstream traceability, the company's shipment date and time and product information must be linked to the identification number of the product to be delivered. In this way, it is now possible to trace back the shipping date, shipper and quantity of each shipped product.

In order to establish traceability, it is important to clearly outline the external flow related to a factory. Taking the example of a factory that manufactures engines, it can be assumed that it acquires components from several suppliers. If one of the components is of poor quality, it can thus easily identify the supplier in question provided that one-tier upstream traceability has been established. If one-tier downstream traceability is also ensured, the company can recall motors manufactured with the components received in the same batch by accurately identifying the recipients in order to minimize losses.

2.2.3.2. *Adding internal traceability information*

The next step is to understand the flow followed by raw materials, parts and products within the factory. In the example mentioned above taken from the automotive industry, internal traceability is ensured by adding and recording before and after each process relevant information such as the manufacturer for each used component (pulley, belt, etc.) and the assembled components (Figure 2.10).

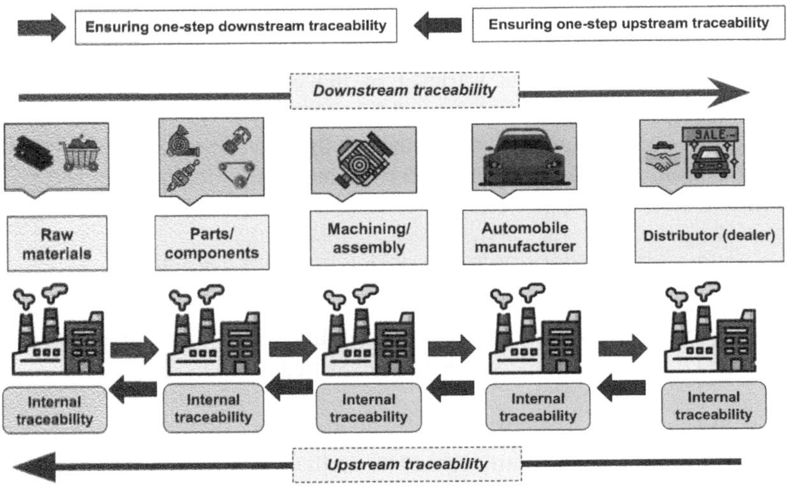

Figure 2.10. *Overview of link management in internal traceability*

The main goal of link management is to link processes one by one while maintaining internal traceability and traceability to one-tier upstream and downstream operators. If the link is established between every process, for example between the supplier of the ventilation belt and radiator hose, the factory responsible for assembling the engine and the car manufacturer that installed the engine on its vehicles, chain traceability is naturally ensured.

2.3. Regulations, laws and standards

This section presents the basic regulations, laws and rules in force for traceability and quality management that can be applied to any industry, namely in the automotive, electronic components, food and pharmaceutical sectors.

2.3.1. *General regulation*

Some standards and laws, such as the International Organization for Standardization (ISO) 9000/9001 series and the Japanese Measurement Act, are widespread in quality management and production control and in the field of traceability.

2.3.1.1. *ISO 9000*

In the ISO 9000 standard, the International Organization for Standardization (ISO) defines traceability as: "The ability to trace history, use, or location of an item or activity through recorded identification". This definition involves two essential prerequisites:

– to clearly specify the identification unit (individual or batch) of the concerned object (raw material, part or product);

– to identify the concerned object and capture the information needed for upstream/downstream traceability

2.3.1.2. *ISO 9001*

The ISO quality management system standards are grouped under the series name ISO 9000; the most important of which is ISO 9001. A quality management system refers to "the part of a management system intended for quality and control management in an organization".

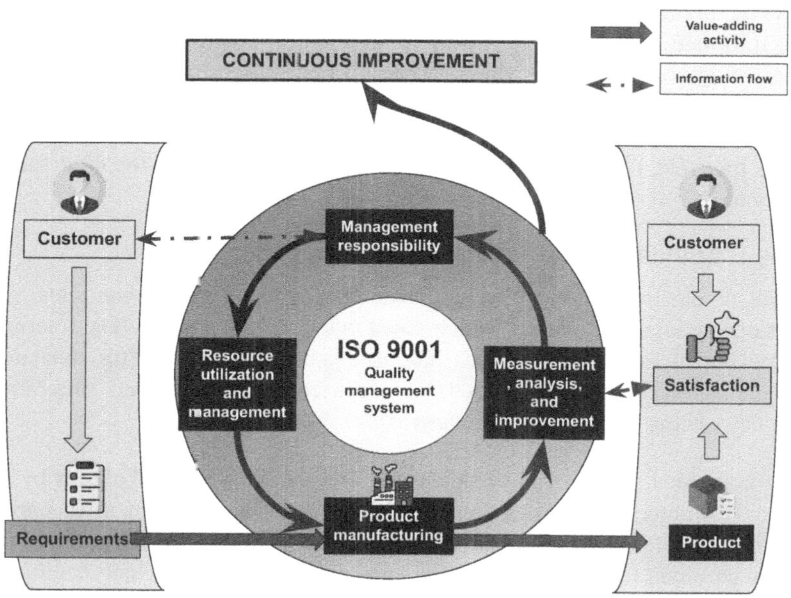

Figure 2.11. *Description of the ISO 9001 standard*

ISO 9001 is an international standard based on several case studies focusing on improvement. Its goal is to maximize profits by guiding companies with a consistent set of rules, applicable not only to manufacturing processes and products, but to the entire chain, from procurement to after-sales service, manufacturing and distribution. It drives process optimization by eliminating problems such as inefficient procedures or repeated errors, which makes it a reference standard.

2.3.1.2.1. Identification and traceability

ISO 9001 specifies the concept of traceability in two separate sections: measurement traceability and identification and traceability. Since this book covers traceability through the entire production chain, we will focus on the identification and traceability section. In terms of identification and traceability, ISO 9001 outlines the following requirements:

– define appropriate means for identification (representation format and method) at each chain stage, from production to distribution;

– make use of the chosen means, identify outputs, such as finished products, and information about each process;

– appropriately record and store the information.

In these requirements, identification is a particularly important concept. Common and shared information can be accessed by all production processes and at any stage of the marketing chain. Consider the example of a chip used to assemble a television. The fundamental principle of identification is to make sure that this chip is equally recognized whether on its own or integrated into a product and to gain knowledge about its lifecycle, that is, its condition, its inspection history and the processes it has undergone.

2.3.1.2.2. Standards for electronic control systems

A lot of electronic control systems can be found on production sites, namely programmable logic controllers, which assist industrial automation. The International Electrotechnical Commission (IEC) has established the IEC 61508 standard as a functional safety standard intended for electrical, electronic and programmable electronic systems common to all industries.

The main idea is to guarantee safety and prevent accidents and claims by adding functional equipment for protecting life, health and the environment (geographical area) from any dangerous hazard. The IEC 61508 standard forms the basis of all currently enforced functional safety standards. The ISO 26262, which is a functional safety standard for automotive electrical/electronic systems, is also derived from the IEC 61508 standard.

2.3.2. *Automotive industry regulation*

This section introduces traceability-related laws and regulations intended for the main players of the automotive industry; these range from car and motorcycle manufacturers, engine and transmission assembly companies and their suppliers of pistons, crankshafts and other parts to providers of navigation systems, tires and other peripheral equipment.

2.3.2.1. *Quality management standard for suppliers*

Manufacturers who produce and market cars or motorcycles are supplied with parts by subcontractors (suppliers). These manufacturers are responsible for overseeing their suppliers' quality management systems. When dozens of suppliers are involved, unified evaluation criteria should essentially be defined.

The international IATF 16949 standard is a step forward in this direction by specifying the quality management guidelines that car and motorcycle manufacturers must impose on their suppliers in order to meet quality, deadlines and other requirements.

2.3.2.2. *Standard for electronic systems*

Modern automobiles are equipped with many electrical/electronic devices, including an electronic control unit (ECU), sensors and actuators (motors). ISO 26262 is an international standard for the functional safety of road vehicles derived from the IEC 61508 standard that applies to industrial equipment and products in general.

It covers the entire lifecycle, from the definition of requirements to disposal, including development, production, maintenance and operation. Consequently, it requires functional safety management covering the whole marketing chain and therefore properly applies to traceability.

Confronted with the rise of electric and hybrid engines, the automotive industry needs to integrate increasingly more electrical/electronic devices. As a result, car manufacturers are constantly looking for suppliers and manufacturers of these ISO 26262-certified devices.

2.3.2.3. *Recall obligation*

Title 49 U.S.C. Chapter 301 – MOTOR VEHICLE SAFETY is a recall system introduced in the United States in the mid-1990s. When a vehicle or one of its pieces of equipment is expected to contain a safety defect or when a new vehicle does not comply with safety standards, the United States recall system requires manufacturers to notify the appropriate authorities and users of the problems or non-conformity. It also imposes a free-of-charge recall or repair obligation for the concerned vehicle models.

Throughout the world, there are laws and regulations governing recalls and other obligations (Table 2.3).

Country	Regulation
United States	Title 49 USC Chapter 301 – MOTOR VEHICLE SAFETY
Japan	Road traffic act
European Union	Directive 2001/95/EC on general product safety
Germany	Gerate und Produktsicherheitsgesetz (product and equipment safety act)
United Kingdom	Code of Practice on Vehicle Safety Defects
Australia	Competition and Consumer Act 2010

Table 2.3. *Overview of recall regulations concerning the automotive industry throughout the world*

2.3.2.4. *A stricter recall system*

Title 49 U.S.C. Chapter 301 – MOTOR VEHICLE SAFETY requires that a recall must be issued in any case of non-compliance with safety standards, even when the cause is unknown. In addition to these already stringent requirements imposed on car and component manufacturers, the United States has further developed its recall system following the tire tread separation problem on Ford Explorer vehicles equipped with Bridgestone/Firestone tires in 2000, which resulted in numerous fatalities and injuries.

The event led to devising and enacting the Transportation Recall Enhancement, Accountability, and Documentation Act (TREAD Act), which came into force in 2000. The TREAD Act, which mandates the compulsory reporting of all safety-related information down to the smallest detail, has suddenly propelled traceability to the forefront of industry quality requirements.

2.3.3. *Regulation of food and pharmaceutical industries*

In addition to international regulations such as ISO 9000/9001 for quality management in the industrial world in general, food and pharmaceutical industries follow specific traceability legislation according to each country.

We will present the laws and regulations applied to food and pharmaceutical industries in some countries, mainly the General Principles of Food Law in the EU and the Bioterrorism Act in the United States.

2.3.3.1. *HACCP*

The HACCP is a food hygiene control method, based on the report published in 1993 by the Codex Alimentarius Commission, a joint organization of the Food and

Agriculture Organization (FAO) and the World Health Organization (WHO) of the United Nations, applied worldwide.

It consists of the scientific analysis of potential food safety risks, such as poisoning risks, in the entire food chain, from the reception of raw materials to shipping, including manufacturing, and processing, and then in identifying critical control points to minimize or eliminate these risks. In Japan, the HACCP was institutionalized through the "Integrated hygiene management in the manufacturing process" amendment, published in 1995.

Within a conventional hygiene inspection framework, the standards applied to the facilities/equipment and food processing methods are specified, and then the final products are submitted to control. The HACCP system recommends performing a risk analysis with each manufacturing process, intensive controls during important processes and control/inspection on the basis of a standard specifically devised for each process. Countermeasures taken along the chain are means of preventing food poisoning and contamination by foreign particles before a problem occurs.

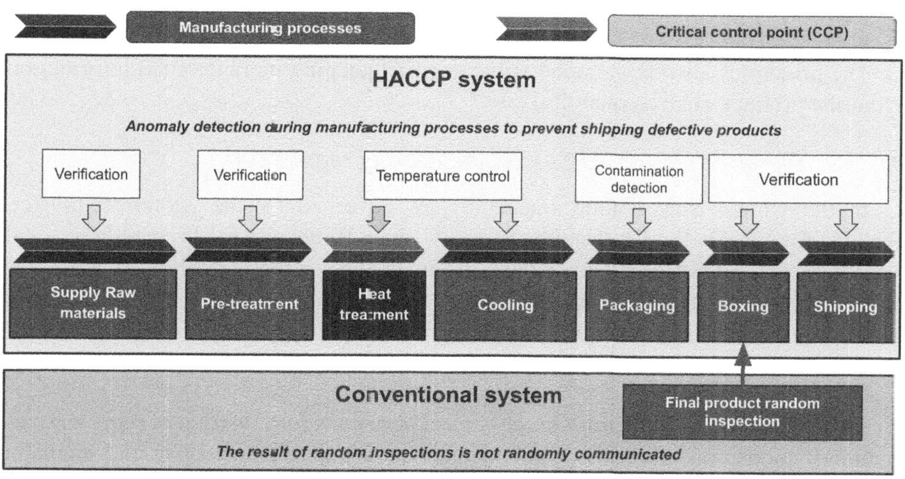

Figure 2.12. *HACCP characterization*

2.3.3.2. *Food traceability principles in the European Union*

Before the surge of mad cow disease (BSE) in Europe in 1996, the European Union did not have unified principles concerning food legislation. At that time, there was only one set of laws that were individually established for each case. This became particularly evident with the BSE crisis, which led to establishing a unified regulation referred to as the general principles and requirements of food law, which was enacted in 2002. The latter lays the foundations for the general principles and requirements of food law, thus leading to the creation of the European Food Safety Authority by Regulation No. 178/2002.

2.3.3.2.1. Specifications in traceability registers

With regard to traceability, Article 18 of the general principles and requirements of food law stipulates the following principle: "the traceability of food, feed, food-producing animals, and any other substance intended to be, or expected to be, incorporated into a food or feed shall be established at all stages of production, processing and distribution".

It is specified that key players in the food chain must make the following information available to the competent authorities at their request:

1) name;
2) place;
3) type of supplied product;
4) transaction and delivery date.

The principles also recommend collecting and keeping more detailed information about the product, such as quantity.

2.3.3.3. *Bioterrorism Act (U.S. food traceability regulation)*

In the United States, food traceability is governed by the Bioterrorism Act, namely the 2002 Public Health Security and Bioterrorism Preparedness and Response Act (PHSBPRA). It dictates the mandatory implementation of measures such as registration of agri-food facilities, prior notice of imports, creation and maintenance of records as well as administrative detention. The traceability-related law measures are linked to record keeping.

Operators who manufacture, process, package, transport, distribute, receive, store or import food products in the United States are required to establish and maintain records. Link management between tier-1 downstream and tier-1 upstream processes, as described in section 2.2.3.1, is required by law. In addition, if a product is suspected of posing a threat to human or animal health, the relevant operators will be required to make all records and other information available to the Food and Drug Administration (FDA).

2.3.3.3.1. Record retention period

The Bioterrorism Act defines the preservation time based on the perishable nature of the food product (Table 2.4).

Category	Non-carriers	Carriers
Products with high degradation risk within a period less than 60 days	6 months	6 months
Products with high degradation risk within a period of 60 days to 6 months	1 year	1 year
Products with high deterioration risk after 6 months	2 years	1 year
Animal feed, including domestic animals	1 year	1 year

Table 2.4. *Definition of traceability record retention periods by the Bioterrorism Act*

2.3.3.4. *Pharmaceutical industry traceability in China*

The medicine management code is a system administered by the China Food and Drug Administration (CFDA), which has been in full force since 2016. Operators wishing to market drugs in China must first obtain a drug management authorization and code from the CFDA and integrate this information into a barcode.

The drug management code was introduced in 2012 for preventing incorrect drug use, implementing traceability and improving distribution. Despite this, its application has so far been limited, all companies, including foreign operators, are required since 2016 to put the system into practice. It is mandatory that the barcode be inserted on a label or directly on the outside of the packaging of each product and, if applicable, of each individual dose.

2.3.3.5. *Summary of traceability standards in several industries*

2.3.3.5.1. Drug management code

The codes comprise 20 numeric characters, coded in Code128C barcode form, of which an example is presented in the following:

$$8 - 123456 - 123456789 - 1234$$

whose elements are thus separated:

1) fixed number, one character;
2) product code, six characters, assigned by the CFDA;
3) serial number assigned to each product;
4) verification code, four characters.

Industrial sector	Standard	Application area	Creation date	Description
	ISO 9000	World	1987 (revised in 1994, 2000, 2008 and 2015)	Standard for a general definition of traceability
General	ISO 9001	World	1987 (revised in 1994, 2000, 2008 and 2015)	Major standard providing a set of rules to improve processes such as traceability
	IEC 61508	World	2002	Functional safety standard for electrical and electronic systems
	IATF 16949	World	2016	Standard providing guidance to car manufacturers about the requirements to be addressed to their suppliers in terms of quality
	ISO 26262	World	2011, 2018	IEC 61508-derived standard for the safety of electronic systems in automotive equipment
	49 US Code Chapter 301 – MOTOR VEHICLE SAFETY	United States	1994	Standard for regulating vehicle recalls in case of non-compliance
	TREAD ACT	United States	2000	Act making the reporting of all vehicle safety-related information mandatory
Vehicle	Road transport vehicle Act	Japan	1960	Set of directives for ensuring road safety (standards, inspection, vehicle maintenance)

Industrial sector	Standard	Application area	Creation date	Description
	Competition and consumer Act 2010	Australia	2011	Trade practices Act
	2001/95/EC directive on general product safety	European Union	2001	Consumer protection directive for controlling the marketing of unsafe products on the European market defining what is a safe product
	Gerate und Produktsicherhe	Germany	2004	Product and equipment safety Act
	Code of Practice on Vehicle Safety Defects	United Kingdom	2019	Code with practices related to safety defects in vehicles
Food and pharmaceutical	HACCP	World	1993	Hygiene control method for identifying potential hazards in manufacturing processes as well as critical control points throughout the entire food chain
	Principles of food Law	European Union	2002	Unified regulation of food law-related principles in the European Union
	Bioterrorism Act	United States	2002	Act governing food traceability principles in the United States such as record retention times, facility registration, etc.
	Drug management code (DEA Code)	China	2012	Drug traceability method in China for product (or single individual dose) identification as well as its marketing authorization number

Table 2.5. *Summary of traceability standards and regulations*

2.4. Status of traceability research

The concept of traceability can be traced back to the 18th century and originated in the food industry (Charlot et al. 2017). To date, traceability is mainly found in this industry, especially for epidemic prevention (Badia-Melis et al. 2015). In national and international legislation, food traceability requirements are enforced to try to mitigate these issues (Olsen and Borit 2018).

The technology found in traceability systems has been used in the fight against product counterfeiting, mainly for drugs, which has become a major issue in the pharmaceutical industry. To this end, unique microtaggants that cannot be copied can be incorporated into the drug. This can be done by tagging the container, the drug capsule, or the particles inside the drug itself. In the medical industry, similar techniques are employed for the monitoring of medical devices and instruments.

In addition to the medical sector, aerospace manufacturing industries have also become increasingly aware of the need for implementing product and process traceability and the benefits that it could bring about. Airbus has been electronically tracing parts using RFID tags since 2009 to track certain parts throughout their lifecycle. Traceability has also been used in tracing machine operations such as CNC (*Computer Numerical Control*) manufacturing using the STEP-NC machine tool control language. By integrating a traceability system into these machines, a larger quantity of data becomes available about the product and its production processes.

2.4.1. *Terminology and definitions*

There are many traceability-related definitions, many of which can be found in the agri-food sector standards and regulations. A modern definition for traceability was proposed by Olsen and Borit (2013) as being "the possibility of accessing all or part of the information relating to what is considered, at every stage of its lifecycle, by means of stored identification information".

A review of the terminology, techniques and technologies related to the implementation of a traceability system is proposed by Schuitemaker and Xu (2020). Although the definitions and terms used in the literature may differ, some concepts systematically emerge such as the notion of Traceable Resource Units (TRU) (Olsen and Borit 2013), which corresponds to the name of the element that is to be tracked.

Traceability can be described as passive or active. The purpose of a passive traceability system is only to provide data visibility, while an active system strives to optimize and better control processes along the supply chain.

Based on the notion of traceable resource unit, the definitions involving upstream traceability and downstream traceability can be more clearly explained: downstream traceability refers to tracking the TRU over time (from creation to destruction), whereas downstream traceability enables tracing back to the origin of the TRU. Upstream and downstream notions are, respectively, referenced by the terms *tracking* and *tracing* (Techsol 2017).

Traceability can be implemented at different completeness levels. The IPC-1782 standard for the electronics industry breaks down traceability into four levels: basic, standard, advanced and comprehensive (Shearon 2018).

The private company GS1, specialist in industrial standards, defines a "traceability level" as the combination between the accuracy of the identification of the TRU and the logistics hierarchy level (GS1 2017). This definition stipulates that the highest traceability level corresponds to a unique and individual TRU identification. For example, a toy supplier might adopt low-level traceability by grouping products by shipment number. Conversely, a medical device supplier would use high-level traceability instead to individually trace products based on their use, but also to ensure proper recycling or disposal at the end of the device lifecycle.

2.4.2. *Traceability data*

A description of the three main components in a traceability system is proposed by Olsen and Borit (2018): identify the traceable resource units (TRU), document their transformations, their connections to each other and record their attributes. Traceable resource units (TRU) can be a business unit (crate, bag, carton, etc.), a logistics unit (pallet) or a production unit (IoT, batch).

Various aspects of traceability data are underlined in Bougdira et al. (2020), such as their heterogeneous nature or interoperability and integration-related challenges. Regulatory and requirement aspects are also included in the study, as they de facto drive traceability policies. All of these aspects are taken into account in Bougdira et al. (2020) and their authors propose an ontology-based modeling description of an intelligent traceability system. In order to evaluate the suggested model, they proposed a cloud-based application whose description focuses on the accuracy and benefits of the system. Another remarkable study can be found in Zhong et al. (2017), which introduces the concept of Smart Manufacturing Objects (SMO) into a traceability platform in order to achieve real-time production within a smart factory. A combination of different technologies such as the IoT (Cohin and Sondi 2015), radio frequency identification (RFID) (Bibi et al. 2017) and laser scanner makes it possible to convert any resource into SMO. In this way, operations and production behavior can be monitored in real time, allowing for detailed and constantly up-to-date traceability. A review of the benefits of traceability regarding supply chain resilience is proposed in Razak et al. (2021), while a comprehensive analysis of traceability cybersecurity in the supply chain is described in Syed et al. (2022).

Most of this research addresses confidentiality in terms of data disclosure, which gives rise to solutions that primarily focus on data protection, excluding them from the traceability data exchanged between partners. In our work, we consider that confidential data should be included, in an appropriate manner, which remains to be defined, in the exchanged traceability data when they can contribute to accelerating retrospective investigations. This subject will be addressed in the next chapter of this book.

2.5. Product-oriented traceability state-of-the-art

2.5.1. *Research on product-oriented traceability*

A product is a set of attributes, called characteristics, devised such that they can be identified (Bettín-Díaz et al. 2018). These characteristics determine the value of what the product is. Product traceability is defined by the ISO 8402 standard as "the ability to trace the history, application or location of an item or activity, or similar items or activities, by means of recorded identification" (Barata et al. 2017). In manufacturing, this means recording the specific movements and changes in a product throughout its lifecycle in the supply chain. A product-oriented ontology developed by Panetto et al. (2012) proposes a methodology that formalizes the technical concepts and embedded data within the product itself, thus promoting interoperability with other applications. The conceptualization of the information model is based on standards such as ISO 10303 and IEC 62264. The field of product traceability proves to be heterogeneous and therefore requires several different technologies (QR code, barcode, RFID), which as such complexifies to a large extent its integration into an already existing system. Within this context, a systematic approach to product traceability was devised by Barata et al. (2017), including an outlook of what is done in the ceramics industry. This approach combines several technologies at different product lifecycle stages and leverages on a multi-level integration of traceability as required by Industry 4.0.

A methodological approach is proposed by Bettín-Díaz et al. (2018) that aims at characterizing products, requirements and key stakeholders involved in a traceability system. In order to track the aggregated spare parts in a manufactured product, Hasan et al. (2020) outlines a blockchain-based approach that addresses some of the traditional problems faced by supply chain management, such as transparency and active monitoring of operations that jeopardize the compliance and safety of spare parts components. Due to its immutability, blockchain technology is not only a means to identify faulty or counterfeit parts, but also to determine the source of actions within the traceability framework. The system proposed in Hasan et al. (2020) includes decentralized file storage based on the Interplanetary File System (IPFS) for spare parts data, smart contract algorithms, security and cost analysis. An assessment is proposed for demonstrating the reliability of the monitoring solution of the properties of spare parts. However, this solution only focuses on interactions occurring within the same company. As a result, the challenges resulting from maintaining critical data confidentiality as well as transparency within the context of interoperability with external partners were not addressed.

Another example of blockchain-based product traceability is introduced by Alkhader et al. (2020) for additive manufacturing. The defined goal was to ensure secure and reliable traceability, as well as accessibility and immutability. The proposed architecture utilizes Ethereum and IPFS smart contracts as a distributed file storage solution. The authors propose an assessment based on security requirements such as integrity, accountability, non-repudiation and authorization. Nonetheless, issues related to critical data management, transparency and trust among stakeholders are not addressed in the study. The cost introduced by using IPFS as a complementary storage solution to blockchain is not analyzed. Concerning the assessments, a survey on blockchain solutions for a sustainable factory and the autonomy of product lifecycle management is proposed in Leng et al. (2020). Various trust-related measures for the use of blockchain solutions in the factory are introduced, involving transparency, decentralized decision-making, reputation and customer relationship. From a factory perspective, blockchain could be used as a catalyst to drive already existing systems for shop floors, such as enterprise resource planning (ERP) and the manufacturing execution system (MES).

From a product management perspective, blockchain would afford a unified database for sharing product information. The fact that new proposals are still being presented in various manufacturing sectors (Tönnissen and Teuteberg 2020), including textiles (Hader et al. 2022), brings forward the importance of practical blockchain solutions for traceability in Industry 4.0. In fact, other technical challenges such as meeting privacy requirements, addressing energy consumption and storage scalability issues are some of the main issues that actually prevent manufacturers from effectively implementing their blockchain solutions in the factory beyond simple proofs of concept (Xu et al. 2022).

2.5.2. *Technological innovations in traceability systems*

Ensuring traceability is now an obligation. In addition to identification by means of labels and marking, every company must manage the links between information and objects.

The rise of traceability is most certainly due to improved computer technologies that today make possible comprehensive information processing by means of marking and reading systems, more accurately than ever before.

Although traceability systems used in industries vary in nature, there is one specific core element that is common to all, and that is the product identification method

(Bai et al. 2017; Schuitemaker and Xu 2020). This identification is generally done by associating the TRU with a unique identifier for distinguishing the physical parts of the product but that is also used for linking these parts with their respective data.

One of the most used techniques for identification purposes is product marking, which can be achieved in two ways: indirect and direct marking (Schuitemaker and Xu 2020).

2.5.2.1. *Indirect marking*

Indirect marking refers to the use of a separate identification carrier, usually a sticker with a number, an alphanumeric code or 2D barcode. This method is usually only employed with larger parts.

The major drawback of labels is that they can be prone to reading errors when they are placed on objects with a non-flat surface (such as a pipe) or attached to specific materials (metallic, brushed and molded).

2.5.2.2. *Direct marking*

Direct part marking (DPM), where serial numbers, lot numbers and 2D codes are directly marked on parts and products, has also largely contributed to the development of traceability systems. In printed circuit board manufacturing sites, fake chips have become a major problem. Direct part marking has been applied as a countermeasure. The combination of direct marking and 2D codes is a method for marking identification symbols on components that are too small for a label to be afixed (Figure 2.13).

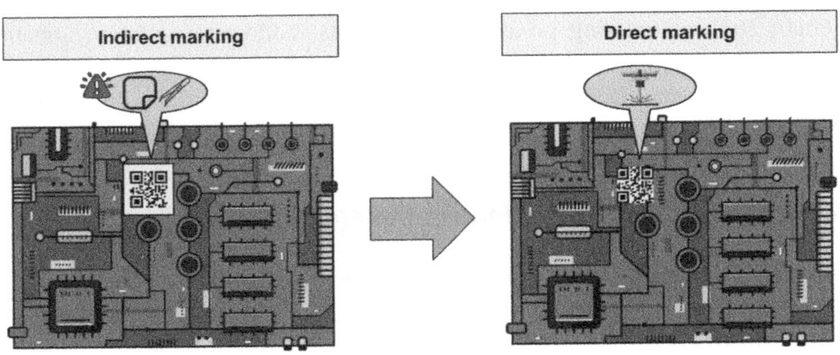

Figure 2.13. *Comparison between indirect and direct product marking*

Direct marking is also suitable for products subjected to high temperatures that operate outdoors for long periods of time, such as on engines and transmissions. It is strongly recommended to avoid the use of labels on these products because they might fall off, risk discoloration or, even more serious, deterioration. Direct marking avoids any problems with reading errors.

2.5.2.3. *Barcodes and information storage*

Traceability requires that a large amount of information be associated with products, namely part numbers, quantity, serial number and shipping volume. Unfortunately, conventional barcodes known as linear (or 1D) (Delcuvellerie and Erben 2020) cannot contain all of this information (Figure 2.14).

Figure 2.14. *Barcode and 2D code information storage*

2D codes are capable of storing more information in a smaller area and are therefore employed in various part and product management processes, including traceability (Óskarsdóttir and Oddsson 2019). These codes can even be marked on very small components, such as miniaturized high-density printed circuit boards, thus allowing for increased individual management reliability.

The most widely used types of 2D codes are the QR code, the datamatrix and PDF417, which have a much larger physical size than the previous two (DynamSoft 2020). The datamatrix is the smallest 2D code with a size of 10 × 10 modules. On the other hand, the QR code measures at least 21 × 21 modules.

The encoding method of these codes is also defined in standards such as the ECC200 standard, which requires the use of the Reed–Solomon algorithm. The advantage of this algorithm is that it reduces the risk of errors when reading a datamatrix code to 1 in 613 million in the best case scenario. In comparison, a standard barcode has a scanning error probability of 1 in 4.5 million at best.

Symbology	Symbol	Type	Creation date	Related standards	Characteristics	Use cases
EAN/UPC	0 01234 56789 5	1D	1970 (EAN), 1976 (UPC)	ISO/IEC 15420:2009	13-character storage (EAN-13 standard most widely used) or 12 for UPC-A, digital data storage only	Existing on almost all consumer products worldwide (UPC in the United States and EAN for international products)
Code 128	123ABC	1D	1981 by Computer Identics Corporation (USA)	ISO/IEC 15417:2007	One of the densest linear symbologies encodes the full 128-character ASCII character set	Shipping tracking
Code 39		1D	1974	ISO/IEC 16388:2007	Less dense than code 128, thus requiring more space on the label	Identification, inventory and tracking
Data Matrix		2D	1994 by Data Matrix. Inc	ISO/IEC 16022	Storage of up to 2,335 alphanumeric characters, ideal for encoding large data volumes in a small space	Pharmaceuticals, PCBs, surgical instruments, manufacturing serial numbers
Aztec		2D	1997 by AIM. Inc	ISO/IEC 24778:2008	Stores up to approximately 3,000 characters, dynamic symbol size based on the amount of data	Transportation tickets, patient ID wristbands, labels for single-use drugs, blood products and samples
QR-code (*Quick Response Code*)		2D	1994 by Denso. Inc	ISO/IEC 18004:2015	Able to store between 4,000 and 7,000 characters	Mobile apps (order confirmation, coupons) and encoding of large data volumes
PDF417		2D	1989 (Symbol Technologies)	ISO 15438	Able to store between 1,850 and 2,725 characters	US driver's license, ID cards, passports, inventories

Table 2.6. *Barcode symbols used for product identification in traceability*

Best practices for the implementation of datamatrix codes have been formulated by GS1 (2018) based on the ECC200 standard. This document is currently used by many industries in the automotive, aeronautics and medical sectors.

The different types of barcodes are called symbologies. A summary of several symbologies is available in Table 2.6.

2.5.2.4. *Optical recognition (OCR) and automation*

OCR is a technology that allows recognition of characters contained in an image. The vision system captures a target image using a camera, recognizes the characters in the captured image and compares their shape to the data stored in the dictionary (fonts) to read them as characters. OCR has become a standard identification technology and for recording expiration dates or part numbers. As a result, errors inherent to reading and manual data entry can also be avoided (Tzoulis and Andreopoulou 2013; Segura Velandia et al. 2016).

Its development has now led to accurately reading and recording characters such as expiration dates and part numbers. As a perfect example of the growing awareness of safety, food expiration dates now not only include an expiration date, but also a production date. The greater the amount of information, the heavier the workload on the production site. Automating the reading process with OCR solves security and workload issues.

OCR, 2D codes (QR code, datamatrix) and direct marking have been adopted by many companies in order to implement traceability. Other new technologies are appearing, such as inkjet printers and radio identification, which are capable of reading/writing information with chip tags and radio waves.

2.5.2.5. *Internet of Things (IoT)*

Within the context of the supply chain, radio frequency identification has been used to identify and track shipments. In the cold chain (refrigeration), data loggers have also been used to monitor shipment transport conditions. In both cases, data visibility was limited and users could not access any shipment data while the transport operation was in progress. There is therefore a real need to use new connected solutions to improve mechanisms for traceability data collection. Integrating IoT into the supply chain improves stakeholder visibility, as well as the tracking and traceability of transported shipments along the supply chain. The purpose of this integration is to improve event detection capabilities, but also to "create meaning" at the level of the collected data (Macaulay and Kückelhaus 2015), all with the aim of improving decision-making. Many recent works recommend the use of this technology, in particular for improving traceability (Guzmán 2018; Li et al. 2019; Arif et al. 2020).

The development of IT technologies promotes coordination between traceability systems, industrial production systems and ERP software, not only benefiting production but every company activity. In addition, information can be shared over a network between the plant and its subsidiaries or partners, independently of their location, whether at the national level or overseas.

The scope of traceability is expected to grow as identification methods diversify and information collection and transfer technologies become more sophisticated.

2.6. Proposal for an approach to traceability for Factory 4.0

This approach was developed during research work for a doctoral thesis involving the authors (Mullet 2022).

2.6.1. *Product perspective*

The concept of intelligent traceability ensures product monitoring in parallel with its tracing and tracking in order to improve its quality and safety (Bougdira et al. 2020). The relevance of product traceability is emphasized in the ISO 9001:2015 quality management standard (Barata et al. 2017). Several technologies and different types of information are involved in the different phases of a product lifecycle, namely development, production, use and disposal. In this work, our approach focuses on traceability with a product manufactured in a supply chain, which may involve components provided by third-party entities such as suppliers. A similar approach was adopted in Hasan et al. (2020) concerning the integration of the different spare parts for building a car engine.

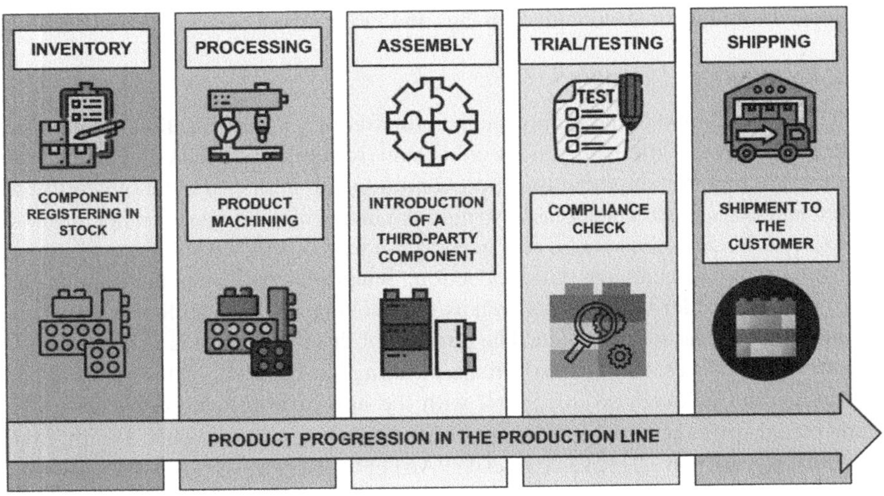

Figure 2.15. *Product progression along the production chain*

Our approach instead focuses on product traceability along the manufacturing process, in which the integration of replacement components is merely one stage. Figure 2.15 illustrates the typical progression of a manufactured product throughout the *supply chain*. This progression involves five main stages, as detailed below.

2.6.1.1. *Inventory*

This is the point where different product components are recorded into the inventory. At this stage, traceability data may consist of component references, serial numbers, or any other information that would help trace products or components.

2.6.1.2. *Process*

This is the stage at which the product enters a production line where it will undergo transformation operations through different sub-stages. At this level, the key process or machine-related parameters are traceability data and their confidentiality level can be critical.

2.6.1.3. *Assembly*

This is when a component manufactured by a third party is integrated into the product. Component-related technical characteristics, how it was used during the assembly process and the impact that including it had on product parameters form the bulk of traceability data at this level. These can be critical for post-assembly auditing.

2.6.1.4. *Test*

This stage consist of verifying the compliance of the product. Traceability data may be any information resulting from the tests (context, measurements, etc.) that could contribute to establishing whether or not a product is compliant.

2.6.1.5. *Shipping*

At this stage, the product is ready to be shipped to the customer. Examples of traceability data are: product batch or shipment number and serial number so that it can be tracked after leaving the factory.

2.6.2. *Traceability data characterization*

According to the ISO 9001 standard, the two objectives of traceability are measurement and identification. For identification, it is requested that appropriate means be used to allow the identification of outputs such as the finished product or the information at the end of each production step. Therefore, a traceability system identifies objects using a representation format (serial numbers, batch numbers) or transfer media (barcode labels, RF tags).

Our approach redefines this normative framework, but we add the aspects described below in order to achieve discriminated data processing in traceability.

2.6.2.1. *Data type*

Traceability data are heterogeneous, and as such they can be of various types. In our approach, we separate them into two categories:

– Simple raw data (RAW) that can be stored in a database. Data likely to be stored in an object database can correspond to this category.

– More complex data such as files that require storage in a file system. These usually tend to occupy more space than simple data.

2.6.2.2. *Confidentiality*

Considering all traceability data, their level of confidentiality has to be known in order to be able to assess security-related risks and costs. Depending on whether these data are stored only in the internal system of the company or whether they need to be transferred over the Internet and stored in the cloud, they will require encryption during transport or permanent encryption, as well as secure storage of the keys that will be used for their subsequent decryption.

2.6.2.3. *Retention*

Retention defines the archival period. This parameter is crucial, especially for the evaluation of data storage-related costs. Indeed, the longer a data record is kept, the longer it will prevent the reuse of already allocated storage space and will therefore contribute to increasing the pressure for acquiring additional storage space for new data.

2.6.3. *Challenges of the approach*

The proposed approach presents a number of challenges and can be described in light of the following principles.

2.6.3.1. *Security*

In the industrial sector, traceability data can be associated with sensitivity or criticality levels. It is therefore important to pay particular attention to this point when deciding which storage medium should be used for traceability data. Losing critical data or finding problems during access is likely to have a significant impact on the business or even the existence of the company; as such, these data must be stored in duplicate form, with at least one copy remaining in the internal system under the full control of the factory.

2.6.3.2. *Transparency*

Generically speaking, transparency is a state in which a system is not suspected of any attempt to hide something. In our approach, the goal is to make it possible for the different entities to be able to corroborate that they all have access to the same information at a given time for all data that individually concern them or that are contractually accessible to them.

2.6.3.3. *Integrity*

Integrity implies maintaining data consistency, accuracy and reliability throughout the data lifecycle, as well as making it possible for entities to be able to verify that this is the case at any time.

2.6.3.4. *Confidentiality*

Confidentiality has been defined by the International Standardization Organization (ISO) as ensuring that information is accessible only to those who have permission. This aspect is particularly important when deciding which data storage means should be used, as mentioned above (section 2.6.2.2).

2.6.3.5. *Non-repudiation*

In the field of information systems security, non-repudiation refers to the possibility of guaranteeing that the confirmed author of a message or action cannot subsequently deny it. This aspect is essential to audit and requires strong authentication of all parties involved in the context that requires non-repudiation.

2.7. Summary of the strengths and weaknesses of the contribution

In this chapter, we have seen the different forms that traceability could take in a factory. First of all, along with the presentation of the foundations of traceability, the various inherent types and notions of this field were introduced. Second, the description of the principles for implementing a traceability system enabled us to show how links and associations between traceability data are established for every manufacturing process of the product. Based on regulations, standards but also on the current state-of-the-art research, it was established that reliable traceability is difficult to achieve, whether in terms of public health in food and pharmaceutical industries, but also in terms of safety, especially in the automotive industry.

In light of these different elements, it is now possible to understand the complex nature of the topic in the sense that product traceability is not limited to a single player, but to a set of key players that, considered as a whole, form what is called the supply chain. The technologies creating these traceability data are heterogeneous and as such the data they generate is also heterogeneous.

In our approach, we focus on product traceability in the factory. Therefore, we divided the process into five steps, and each one according to its needs might involve external stakeholders. For example, it should be noted that in addition to the factory, assembly involves a supplier and shipping involves a customer. A summary showing the accumulation of traceability data during the manufacturing process of the product is presented in Figure 2.16.

Besides this classification in various stages, we assign tags to the different data to qualify them according to various parameters:

- type;
- confidentiality;

– perimeter (single/multi-partner);

– retention.

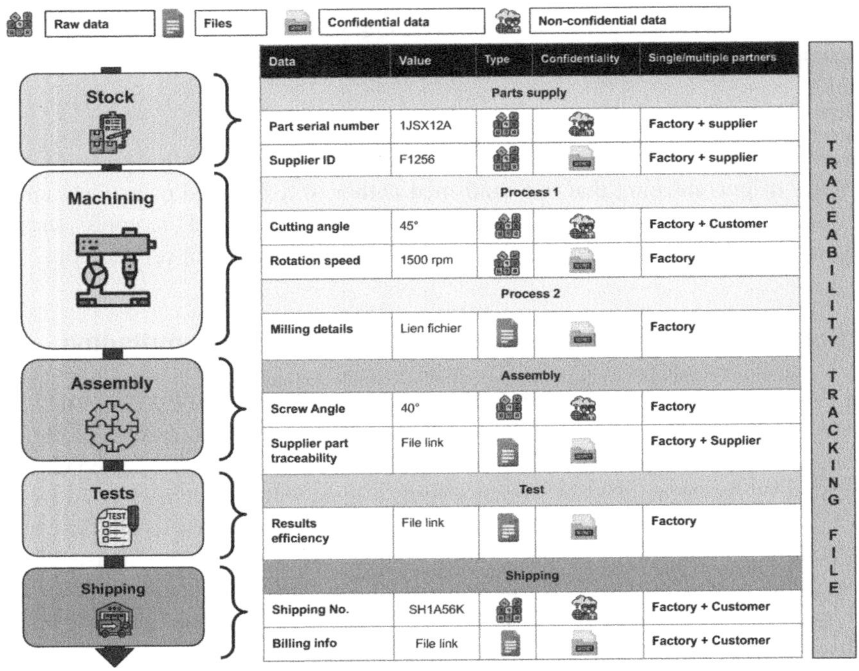

Figure 2.16. *Overview of a product traceability sheet*

The interest of these tags is to get a better and clearer view about the data stored in the factory traceability system in order to simplify the decision-making process in relation to choices about the location where data should be stored (local/cloud) to the precautions that should be taken regarding data security or more generally regarding the technological decisions related to the traceability system.

Given that factories are becoming increasingly more digital, there is a growing demand for enforcing guarantees about the reliability of critical product data, primarily between different factories along the production chain. Recently, blockchain (the technology behind Bitcoin) has been utilized as a way to guarantee that the large amounts of data generated by traceability systems are tamper-proof (Krima et al. 2019). Blockchain is a promising technology that could, jointly with our approach, meet demand by ensuring transparency, enabling all stakeholders to reliably verify the entire history of a product lifecycle.

2.8. References

Alkhader, W., Alkaabi, N., Salah, K., Jayaraman, R., Arshad, J., Omar, M. (2020). Blockchain-based traceability and management for additive manufacturing. *IEEE Access*, 8, 188363–188377.

Arif, J., Jawab, F., Mouzouna, Y. (2020). Design on improvement of traceability process in the outsourcing of logistics' activities using the Internet of Things (IoT) applications. *Maejo International Journal of Science and Technology*, 29, 1093–1108.

Badia-Melis, R., Mishra, P., Ruiz-García, L. (2015). Food traceability: New trends and recent advances. A review. *Food Control*, 57, 393–401 [Online]. Available at: https://www.sciencedirect.com/science/article/pii/S0956713515002698.

Bai, H., Zhou, G., Hu, Y., Sun, A., Xu, X., Liu, X., Lu, C. (2017). Traceability technologies for farm animals and their products in China. *Food Control*, 79, 35–43 [Online]. Available at: https://www.sciencedirect.com/science/article/pii/ S0956713517300865.

Barata, J., da Cunha, P., Gonnagar, A., Mendes, M. (2017). A systematic approach to design product traceability in Industry 4.0: Insights from the ceramic industry. In *Information Systems Development: Advances in Methods, Tools and Management – Proceedings of the 26th International Conference on Information Systems Development*, Paspallis, N., Raspopoulos, M., Barry, C. et al. (eds). ISD, Larnaca.

Bettín-Díaz, R., Rojas, A.E., Mejía-Moncayo, C. (2018). Methodological approach to the definition of a blockchain system for the food industry supply chain traceability. In *Computational Science and Its Applications – ICCSA 2018*, Gervasi, O., Murgante, B., Misra, S., et al. (eds). Springer, Berlin.

Bibi, F., Guillaume, C., Gontard, N., Sorli, B. (2017). A review: RFID technology having sensing aptitudes for food industry and their contribution to tracking and monitoring of food products. *Trends in Food Science And Technology*, 62, 91–103.

Bougdira, A., Ismail, A., Ahaitouf, A. (2020). A traceability proposal for Industry 4.0. *Journal of Ambient Intelligence and Humanized Computing*, 11, 3355–3369.

Charlot, B., Montet, D., Ray, R. (2017). *Food Traceability and Authenticity*. CRC Press, New York.

Cohin, O. and Sondi, P. (2015). Internet of things for smart factory. *IEEE COMSOC MMTC E-Letter*, 10(5), 21–23.

Delcuvellerie, J.-L. and Erben, S. (2020). *Revisiter la traçabilité avec les technologies industrie 4.0*. Editions T.I., Saint-Denis.

DynamSoft (2020). The comprehensive guide to 1D and 2D barcodes [Online]. Available at: https://www.dynamsoft.com/blog/insights/the-comprehensive-guide-to-1d-and-2d-barcodes/.

GS1 (2017). GS1 Global Traceability Standard [Online]. Available at: https://www.gs1.org/sites/default/files/docs/traceability/GS1_Global_Traceability_Standard_i2.pdf.

GS1 (2018). Overview and technical introduction to the use of GS1 datamatrix [Online]. Available at: https://www.gs1.org/docs/barcodes/GS1_DataMatrix_Guideline.pdf.

Guzmán, L. (2018). Sigfox and Louis Vuitton partner for innovative luggage tracker [Online]. Available at: https://www.sigfox.com/en/news/sigfox-and-louis-vuitton-partner-innovative-luggage-tracker.

Hader, M., Tchoffa, D., Mhamedi, A.E., Ghodous, P., Dolgui, A., Abouabdellah, A. (2022). Applying integrated blockchain and big data technologies to improve supply chain traceability and information sharing in the textile sector. *Journal of Industrial Information Integration*, 28, 100345 [Online]. Available at: https://www.sciencedirect.com/science/article/pii/S2452414X22000176.

Hasan, H.R., Salah, K., Jayaraman, R., Ahmad, R.W., Yaqoob, I., Omar, M. (2020). Blockchain-based solution for the traceability of spare parts in manufacturing. *IEEE Access*, 8, 100308–100322.

Keyence (2024). Les bases de la traçabilité [Online]. Available at: https://www.keyence.fr/ss/products/marking/traceability/basicabout.jsp.

Krima, S., Jr. and Feeney, A.B. (2019). Securing the digital threat for smart manufacturing: A reference model for blockchain-based product data traceability. Report, NIST Publications [Online]. Available at: https://tsapps.nist.gov/publication/getpdf.cfm?pubid= 926019.

Leng, J., Ruan, G., Jiang, P., Xu, K., Liu, Q., Zhou, X., Liu, C. (2020). Blockchain-empowered sustainable manufacturing and product lifecycle management in Industry 4.0: A survey. *Renewable and Sustainable Energy Reviews*, 132, 110112 [Online]. Available at: https://www.sciencedirect.com/science/article/pii/S1364032120304032.

Li, X., Yang, L., Duan, Y., Wu, Z., Xiaoshuan, Z. (2019). Developing a real-time monitoring traceability system for cold chain of tricholoma matsutake. *Electronics*, 8, 423.

Macaulay, J. and Kückelhaus, M. (2015). Internet of things in logistics. Technical Report, DHL [Online]. Available at: https://discover.dhl.com/content/dam/dhl/downloads/interim/full/dhl-trend-report-internet-of-things.pdf.

Mullet, V. (2022). Intégration de solution blockchain dans un système global de traçabilité d'une usine opérationnelle. PhD Thesis, Université du Littoral Cote d'Opale, Dunkirk [Online]. Available at: https://theses.hal.science/ tel-04053446.

Olsen, P. and Borit, M. (2013). How to define traceability. *Trends in Food Science and Technology*, 29(2), 142–150 [Online]. Available at: https://www.sciencedirect.com/science/article/pii/S0924224412002117.

Olsen, P. and Borit, M. (2018). The components of a food traceability system. *Trends in Food Science and Technology*, 77, 143–149 [Online]. Available at: https://www.sciencedirect.com/science/article/pii/S0924224417304107.

Óskarsdóttir, K. and Oddsson, G.V. (2019). Towards a decision support framework for technologies used in cold supply chain traceability. *Journal of Food Engineering*, 240, 153–159 [Online]. Available at: https://www.sciencedirect.com/science/article/pii/S0260877418303029.

Panetto, H., Dassisti, M., Tursi, A. (2012). Onto-pdm: Product-driven ontology for product data management interoperability within manufacturing process environment. *Advanced Engineering Informatics*, 26(2), 334–348 [Online]. Available at: https://www.sciencedirect.com/science/ article/pii/S1474034611001121.

Razak, G.M., Hendry, L.C., Stevenson, M. (2021). Supply chain traceability: A review of the benefits and its relationship with supply chain resilience. *Production Planning & Control*, 34(11), 1114–1134.

Scantrust (2024). Guide sur la traçabilité des produits dans les chaînes de distribution [Online]. Available at: https://www.scantrust.fr/guide-sur-la-tracabilite-des-produits-dans-les-chaines-de-distribution/.

Schuitemaker, R. and Xu, X. (2020). Product traceability in manufacturing: A technical review. *Procedia CIRP*, 93, 700–705 [Online]. Available at: https://www.sciencedirect.com/science/article/pii/S2212827120306922.

Segura Velandia, D.M., Kaur, N., Whittow, W.G., Conway, P.P., West, A.A. (2016). Towards industrial internet of things: Crankshaft monitoring, traceability and tracking using RFID. *Robotics and Computer-Integrated Manufacturing*, 41, 66–77 [Online]. Available at: https://www.sciencedirect.com/science/article/ pii/S0736584516300576.

Shearon, C. (2018). IPC-1782 standard for traceability of critical items based on risk. In *Pan Pacific Microelectronics Symposium*. IEEE, Big Island, HI, 1–3.

Syed, N.F., Shah, S.W., Trujillo-Rasua, R., Doss, R. (2022). Traceability in supply chains: A cyber security analysis. *Computers & Security*, 112, 102536 [Online]. Available at: https://www.sciencedirect.com/science/article/pii/ S0167404821003606.

Techsol (2017). Industry 4.0: The importance of traceability [Online]. Available at: https://medium.com/@Techsol_srl/industry-4-0-the-importance-of-traceability-c9c05691676f.

Tönnissen, S. and Teuteberg, F. (2020). Analysing the impact of blockchain-technology for operations and supply chain management: An explanatory model drawn from multiple case studies. *International Journal of Information Management*, 52, 101953.

Tzoulis, I. and Andreopoulou, Z. (2013). Emerging traceability technologies as a tool for quality wood trade. *Procedia Technology*, 8, 606–611 [Online]. Available at: https://www.sciencedirect.com/science/article/pii/S2212017313001497.

Xu, X., Tatge, L., Xu, X., Liu, Y. (2022). Blockchain applications in the supply chain management in German automotive industry. *Production Planning & Control*, 1–15 [Online]. Available at: https://doi.org/10.1080/ 09537287.2022.2044073.

Zhong, R.Y., Xu, X., Wang, L. (2017). IoT-enabled smart factory visibility and traceability using laser-scanners. *Procedia Manufacturing*, 10, 1–14 [Online]. Available at: http://www.sciencedirect.com/science/article/pii/ S235197891730286X.

3

Blockchain Integration in Traceability for Factory 4.0

Industry 4.0 introduces major changes in the management of manufacturing processes. Third parties, such as suppliers, customers and other providers, are more likely to interact with a factory traceability system to achieve full transparency about the manufacturing process of a product. Several solutions, including blockchain-based approaches, have been proposed in order to foster trust in such systems. However, many plant owners are still reluctant about deploying such solutions, especially due to threats to factory data privacy that could be inserted in traceability data.

This chapter outlines our approach to including critical factory data in a blockchain-based traceability system and to transaction validation associated with third parties without jeopardizing privacy. We first present the main features of the blockchain paradigm as well as the MultiChain platform that we used in our work. This is then followed by a depiction of how the state of the art in blockchain technology can be used in an industrial environment and a detailed description of our proposal for a traceability system based on this technology, as well as its implementation with MultiChain. In addition to privacy protection properties, we show how other challenging problems such as power consumption and storage space could be controlled through the functionalities offered by MultiChain.

3.1. Blockchain paradigm presentation

Blockchain represents a major technological innovation primarily defined as a technology for information storage and transmission. It enables high transparency and

For a color version of all the figures in this chapter, see: www.iste.co.uk/sondimullet/cybersecurity.zip

Cybersecurity and Traceability in Factory 4.0,
coordinated by Patrick SONDI and Valentin MULLET.

security standards because it operates without a centralized control body. In practical terms, connected through a network, its users are allowed to share data without a management or trusted intermediate.

The first blockchain was initiated in 2008 by Nakamoto (2009) and led to the blockchain Bitcoin network being implemented (Bodkhe et al. 2020), which is used in the banking sector and relies on cryptocurrencies.

Although historically Bitcoin was developed to support transactions carried out through cryptocurrencies, its scope for application proves to be much broader and presents many advantages for the industrial and logistics sectors; this is particularly the case for ensuring product traceability and to establish the background history of the various processes involved in a production and distribution chain.

In order to better understand the vocabulary inherent to blockchain, Figure 3.1 is proposed as a supporting reference for the detailed definitions provided through this section.

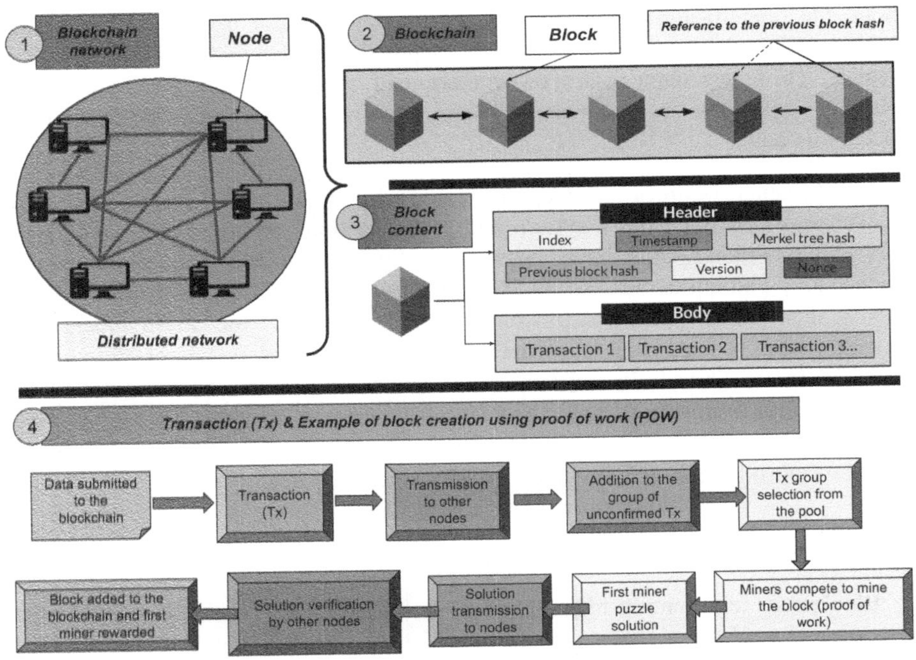

Figure 3.1. *Overview of the main blockchain-related concepts*

3.1.1. *Blockchain*

The blockchain is a distributed ledger that contains the history of all exchanges made between users since its creation. It consists of a growing list of blocks in which each block is linked to the previous one by a cryptographic process. This link between blocks was introduced in 1991 by Bodkhe et al. (2020) and the concept of chain was derived therefrom. The key point in linking is that each block contains the hash of the previous block, which prevents data from being modified after adding them, thus guaranteeing the principle of blockchain immutability (Alkhader et al. 2020). This fundamental characteristic ensures that a block cannot be retroactively altered without the set of subsequent blocks having to be modified. Such a modification would unveil an attempt to tamper with the data.

3.1.2. *Blockchain network*

This term refers to the network formed by all the machines running the blockchain. Each member of this network is called a "node". This is a distributed network also called a peer-to-peer (P2P) network, which means that it does not depend on any central authority to be managed.

3.1.3. *Node*

A node is a machine within the blockchain network. In practical terms, it can therefore be considered that a node represents a part, a member or a user. Each machine is identified through one or more unique addresses whose generation is based on a cryptographic process. These addresses are therefore the guarantee that the authors of the various actions performed in the blockchain can be traced with certainty. Each node locally hosts a copy of the blockchain according to predefined conditions.

3.1.4. *Block*

The block is the first level of decomposition in the underlying blockchain structure. It is divided into two distinct parts: the header and the body. The header contains the index (block position within the chain), the timestamp (Leng et al. 2020) that enables time-related information to be managed, the Merkel tree hash that corresponds to the hash of all the transactions contained in the block, the hash of the previous block thus creating the link between blocks, the version of the cryptographic protocol used, as well as a value called the nonce; the nonce is a value representing the solution of the consensus algorithm whose definition will be proposed later in the text. The block body contains a list of transactions. A block hash corresponds to the hash of all the information contained in its header.

3.1.5. *Transaction*

The transaction is the second decomposition level of the blockchain. It represents a data unit submitted to and stored in the blockchain. Transactions are recorded in the blocks, the number of which is defined by a parameter limiting their total number in a block. When a user submits data to the blockchain, it generates a transaction as well as the corresponding hash that will serve as unique identifier in the blockchain.

3.1.6. *Consensus algorithm*

Since it is a distributed network, the blockchain is not governed by a central authority that ensures that each node has the same blockchain version at all times. This function of ensuring reliability and trust between the various parties of the blockchain network is precisely ensured by the consensus algorithm. Several objectives might be pursued such as collectively deciding whether a transaction is valid, synchronizing machine states to ensure their consistency or designating a node responsible for the execution of a specific task. Consensus algorithms work on the basis that certain processes and systems will be unavailable, as such they must be fault tolerant. According to this assumption, only a number of nodes will respond, and among these a minimum number of responses are required, for example 51%. There are various existing algorithms, including the popular proof of work (POW) found in Bitcoin (Nakamoto 2009).

3.1.7. Smart contract

A smart contract refers to the hard-coding of all the contractual clauses into hardware or software to be automatically executed within a secure and distributed environment (Szabo 1997). In the context of blockchain, this designates computer code stored in the blockchain and defines the transactional rules between the different parties. One of the first implementations of *smarts contracts* in blockchain was proposed by Ethereum in 2013 (Buterin 2014). Smart contracts automate the execution of interactions between the involved parties without the need for a trusted intermediate. By expressing legal relationships in code form rather than in words, they act as a guarantee that transactions can take place directly without any error, whether deliberate or not. These rules can be expressed, executed and validated in many ways in existing blockchain platforms. A comparison of *smart contract* implementations with different blockchain technologies is proposed in section 3.2.4.

3.1.8. *Types of implementations and environments*

A large number of implementations of the blockchain paradigm can be found and these can be grouped into different categories. Some blockchains, such as Bitcoin, are implemented without an authorization system while others are known to be authorized, such as Hyperledger.

With permissionless blockchains, no control is imposed on the access to transactions, while permissioned blockchains have built-in mechanisms for controlling read/write access to transactions submitted by stakeholders; nonetheless this access is authorized to a limited and known number of users.

The environments in which these blockchain implementations can be deployed are classified into three categories: public, private or consortium (Buterin 2015). The choice of blockchain implementation therefore depends on the requirements of the stakeholders.

3.1.8.1. *Public blockchain*

Read and write access on the blockchain is granted to everyone. Any user in the world can join the network, as wells as read and write transactions on the blockchain. This environment also guarantees user anonymity. This type of deployment is best suited to Business to Customer (B2C) or Customer to Customer (C2C) models.

3.1.8.2. *Consortium blockchain*

A list of pre-selected nodes control read and write access on the blockchain. It is useful in environments involving several organizations requiring secure data to be shared, such as in the context of the Business to Business (B2B) supply chain.

3.1.8.3. *Private blockchain*

A single organization controls every access right to the blockchain.

3.1.9. *Examples of consensus algorithms*

3.1.9.1. *Proof of Work (POW)*

This is a consensus algorithm used for the miner selection when appending the next block to the blockchain. The concept consists of making the nodes compete with one another in order to solve a complex mathematical problem called a puzzle. The solution to this problem is called the nonce and was mentioned earlier in the block definition. In this competition, the nodes are referred to as miners. A key feature of the POW concept is the asymmetry related to implementation costs. Furthermore, the work must be hard for the author of the request, whereas it should be easily verifiable by the server, called the verifier. The first node able to solve the puzzle submits its solution to the network, which will perform the verification. If the outcome is valid, the winning miner earns a reward (such as money in the case of Bitcoin). The purpose of this reward is to foster compliance with the rules since there is no point in breaking them. POW also contributes to strengthening security since it makes it difficult to alter data in the blockchain. Any block data alteration would require a posteriori remining all the previously generated blocks and would imply that a large amount of power be consumed (equal to what the whole network had to consume to mine the blocks in question) (Masood and Faridi 2018).

3.1.9.2. *Proof of Stake (POS)*

To solve the problem of energy inefficiency in POW, proof of stake was first introduced by King and Nadal (2022). It proposes linking mining power to the percentage of tokens held by the miner. This significantly reduces power consumption and improves the security of the consensus mechanism.

3.1.9.3. *Practical Byzantine Fault Tolerance (PBFT)*

With this consensus mechanism, a node broadcasts a message (Castro and Liskov 1999). When it receives $2F + 1$ identical responses from other nodes where F is the maximum number of tolerated Byzantine nodes, that is, one-third of the nodes existing in the PBFT network, then the response value is considered to be valid. Consequently, agreement is reached faster than with POW, with less power consumption. PBFT is widely used by permissioned blockchains such as Hyperledger Fabric (Androulaki et al. 2018).

3.1.9.4. *Replicated And Fault Tolerant (RAFT)*

It is divided into two main stages. The first is leader election and the second is log replication. During the leader election stage, a node is selected to operate as the leader and in the event of failure, a new leader is elected. During the log replication stage, the leader accepts client commands, replicates its logs to the other network nodes and when a conflict occurs, the content of the node is overwritten with the leader's logs. This algorithm ensures high transaction throughput. However, it was not designed to manage Byzantine faulty nodes. RAFT is supported by some of the permissioned blockchains such as Hyperledger Fabric (Androulaki et al. 2018).

3.1.9.5. *Proof of Elapsed Time (PoET)*

It was proposed by Intel for solving the problem of faulty Byzantine nodes in the blockchain network. It relies on the use of a Trusted Execution Environment (TEE) to select the miner that will generate the next block. A function is executed in the TEE generating random timer values for the nodes involved, and the node with the lowest timer value is designated for the next block generation. This consequently reduces the amount of computation required, but this algorithm is dependent on the TEE such as the Intel Software Guard Extensions (SGX) implementation, which is restrictive when it comes to executing the algorithm on a network with different hardware architectures. This algorithm is used by Hyperledger Sawtooth (Olson et al. 2018), a permissioned blockchain implemented by Intel.

3.2. MultiChain blockchain platform

Our choice was to present the approach proposed in this book with an *open source* blockchain platform so that it can be accessed by as many people as possible and because it enables us to more openly show how the different choices are made and implemented.

MultiChain is an *open source* blockchain platform for building and deploying private blockchain applications that can be used within the same company or between multiple partner companies. This is a *bitcoin-core* fork, which is at the heart of Bitcoin technology. However, various features have allowed MultiChain to differentiate from Bitcoin ranging from its nature as a private blockchain to the data storage system, not forgetting the mining algorithm which is different from POW.

Not only does it provide a command-line interface, but also a simple API so that it can be integrated into many programming languages. Many features are provided including permission management, the creation of different configurations for each blockchain, as well as data streams.

3.2.1. *Features*

This section aims to introduce some key points regarding MultiChain features.

3.2.1.1. *Permissioned*

As with any private blockchain, access to a MultiChain blockchain network first requires login permission for any new members. This access is given by the blockchain initiating node.

3.2.1.2. *Stream*

The concept of stream enables the blockchain to be used as a private network-shared database. It is in general used for data storage, timestamping, archiving and accessing stored data. In structure terms, this is a collection of objects called *stream items* where each object contains a timestamp, one or more keys for recovery, a digital signature and of course the stored data. Just as with any MultiChain element, a permission system is implemented to designate who can and cannot read and write to a specific stream.

3.2.1.3. *Scalability*

MultiChain can solve scalability issues in terms of network resource and storage usage based on a dual storage mode, which consists of storing data in *onchain* or *offchain* mode. In contrast to other blockchains where transaction content is stored on all nodes, MultiChain tolerates a partial data replication on nodes. This results in a reduced block size, since a block only includes the data hash in *offchain* mode and in *onchain* mode when the data size is too large.

3.2.1.4. *Mining*

Mining, which is the process of creating new blocks, is performed by a group of entities defined and authorized by the administrator(s) of the blockchain network.

3.2.2. *Main parameters*

With the tools integrated by default in MultiChain, many blockchain parameters are allowed to be controlled such as maximum block size, the operation of the consensus algorithm as well as network permissions (Table 3.1).

Parameter	Description	Default value
target-block-time	Average number of seconds between block mining. This is a target value that the blockchain seeks to approximate as much as possible	15 s
maximum-block-size	Maximal number of bytes in each block. This limit is to avoid overloading the network with blocks too large	8,388,608 (8 mB)
maximum-chunk-size	Maximal chunk size when storing data in *off-chain* mode. The lower the value, the larger the number of chunks the files will be split into	1,048,576 (1 mB)
mining-diversity	Minimum proportion of miners required for mining a block. If the proportion of connected miners is lower, then no additional blocks will be mined and the blockchain will therefore be blocked	0.5
mine-empty-rounds	Number of rounds after which block mining is stopped if no new transactions are submitted to the blockchain. This results in a reduction of the blockchain disk usage in low activity periods	10
mining-turnover	A value of 0.0 implies pure mining such as round-robin where nodes mine in turns, while a value of 1.0 implies random block generation between nodes with mining permission	0.5
target-adjust-freq	Frequency at which the mining difficulty is recomputed when using proof of work. If the value is less than 0, then the algorithm used is round-robin	–1
pow-minimum-bits	Initial value related to mining difficulties when using proof of work. This is the number of zeros required upstream in the block hash value	8

Table 3.1. *Main MultiChain parameters*

3.2.3. *MultiChain JSON API*

In order to make it possible to integrate it into various programming languages, MultiChain provides a JSON-RPC API to interact with the blockchain.

The API is implemented as a set of functions, several examples of which are given in Table 3.2.

Function	Parameters	Description
getblockchainparams	N/A	Enables the return of the parameter list to the blockchain
createkeypairs	N/A	Enables the creation of a public/private key pair that can be used by a node. This function could be used by a "cold node" to assign keys to other nodes
publish	streamname, keys, data	Enables the publication of an object in a given stream with one or more keys that will be used for its later retrieval. This function returns the hash of the created transaction
getstreamitem	txid	Enables the object to be retrieved from a stream with the transaction hash value *txid*. This function cannot return more than one result because the transaction hash is a single value
liststreamkeyitems	streamname, keys	Used to list objects in a stream with the key passed as parameter

Table 3.2. *Examples of MultiChain API commands*

These can be called by way of an HTTP POST request to the API listening port. The listing snippet in 3.1 shows the request body sent to the API to publish an object in a stream called *stream1*.

3.2.4. *Comparison with other blockchain solutions*

Most blockchain discussions always make mention of the notion of *smart contracts*. It is a popular belief that smart contracts automate the execution of interactions between parties, without the need for a trusted intermediary. By expressing legal relationships in code form rather than words, it is assumed that they allow transactions to directly take place without errors, deliberately or not.

From a technical standpoint, a smart contract is something more specific, namely computer code stored in a blockchain which defines the transactional rules of this chain. This description appears to be rather simplistic, but a closer look reveals great variability in how these rules are expressed, executed and validated. When deciding which blockchain platform should be used for a new application, the question "Does this platform support smart contracts?" is not the right one that should be asked. Instead, we should inquire: "What kind of smart contracts does this platform support?"

```
{
    "jsonrpc": "2.0",
    "method": "publish",
    "params": ["stream1", "TRACEA_STEP", {"json": {
        "location":"Line9-u-3",
        "process_desc":"Reduced discrete Internet
        ↪  solution"}}],
    "id": 1,
    "chain_name": "MyBlockchain"
}
```

Listing 3.1. *Example of a JSON request to the MultiChain API*

The goal of this section is to look into some of the key differences between smart contract approaches and the trade-offs they represent. Four popular enterprise blockchain platforms that implement on-chain custom code will be compared. First, IBM's Hyperledger Fabric that calls "chaincode" to its contracts. Second, the MultiChain platform, which introduces *smart filters*. Third, Ethereum (and its Quorum and Burrow permissioned spin-offs) that popularized the "smart contract". And finally, R3 Corda, which refers to "contracts" in its transactions. Despite all the different terminologies, they ultimately all refer to the same thing: application-specific code that defines the rules of a chain.

3.2.4.1. *Basic concepts*

We consider an application that is shared across various organizations based on an underlying database. In a traditional centralized architecture, this database is hosted and administered by a single host that all participants trust, even if they do not trust each other. Transactions that modify the database are initiated only by applications on the systems in this central entity, mostly as responses to messages received from participants. The database simply does what it is told because the application is implicitly trusted to send only the transactions that make sense.

Blockchains are an alternative way to manage a shared database, without a trusted intermediary. In a blockchain, each party involved operates a node that contains a copy of the database and independently processes transactions that modify it. Participants are identified on the basis of public keys or addresses, each of which has a corresponding private key that only the identity owner knows. While transactions can be created by any node, they are "digitally signed" by the private key of their initiator so that they can prove their origin.

Nodes directly connect to one another (peer-to-peer), quickly propagating transactions and the "blocks" in which they are time-stamped and confirmed across the network. The blockchain itself is literally a chain of these blocks, which forms an ordered log of historicized transactions. A "consensus algorithm" is used to ensure that all nodes reach agreement on the blockchain contents, without requiring centralized control.

In theory, any distributed database application could be structured using a blockchain at its core. However, this creates a number of technical challenges that a centralized scenario does not have to deal with.

3.2.4.1.1. Transaction rules

If any participant is able to directly modify the database, how can we ensure that they do not break the application rules? What prevents a user from knowingly corrupting the database contents?

3.2.4.1.2. Determinism

Once these rules are defined, they will be applied several times by various nodes when processing transactions for their own copy of the database. How can we be sure that each node receives exactly the same result as the others and thus ensure consistency in the shared database?

3.2.4.1.3. Conflict prevention

With no central coordination, how should two transactions that follow the application rules, but that nevertheless are in conflict with each other be processed? Conflicts may originate from a deliberate attempt to fool the system, or can innocently result from bad luck and timing.

Therefore, where do smart contracts, smart filters and blockchain code come from? Their main goal is to operate with the underlying blockchain infrastructure to solve these challenges. Smart contracts are the decentralized equivalent of application code: instead of running from a central location, they run on multiple blockchain nodes, creating or validating transactions that modify the contents of that database.

Starting form the first of these challenges, the transaction rules, let us observe how they are expressed in Fabric, MultiChain, Ethereum and Corda, respectively.

3.2.4.2. *Transaction rules*

Transaction rules perform a specific function in blockchain-powered databases, restricting the transformations that can be achieved with the database state. This is necessary because blockchain transactions can be initiated by any of its participants, bearing in mind that these participants do not fully trust each other to allow the database to be modified at will.

Consider two situations that exemplify why transaction rules are necessary. First, imagine a blockchain designed to aggregate and timestamp PDF documents published by its participants. In this scenario, no one should have the right to delete or modify documents, as this would compromise the whole purpose of the system – document persistence. Second, consider a blockchain representing a shared financial ledger, which keeps track of users' balances. We cannot allow participants to arbitrarily inflate their own balance or reduce the balance of others.

3.2.4.2.1. Inputs and outputs

Two main approaches form the basis of blockchain platforms to express transaction rules. The first, called the input–output model, is used in MultiChain and Corda. Here, transactions are explicitly registered in the rows or database "states" they delete and create, respectively, forming a set of "inputs" and "outputs". Modifying a row is expressed as the equivalent operation that consists of deleting that row and creating a new one as a substitute.

Given that the database rows are deleted in the inputs only and created only in the outputs, each input must "jump" the output of a previous transaction. The current database state is defined as the set of "unspent transaction outputs" or "UTXOs", that is the outputs of previous transactions that have not yet been used. Transactions may also contain additional information, called "metadata", "orders" or "attachments", which are not part of the database but help define their meaning or purpose.

Given these three sets of inputs, outputs and metadata, the validity of a transaction in MultiChain or Corda is defined by code that can perform arbitrary computations with these sets. This code can validate the transaction, or return an error with a corresponding explanation. The I/O model can be considered as an automated "inspector" that maintains a checklist ensuring that transactions meet each of the rules. If the transaction fails any of these checks, it will be automatically rejected by all nodes in the network.

It should be observed that despite the shared I/O model, MultiChain and Corda implement it in very different ways. In MultiChain, outputs can contain assets and/or data in JSON, text or binary format. Rules are defined in "transaction filters" or "stream filters", which can be defined to check all transactions, or only those involving specific assets or groups of data. In contrast, a Corda output "state" is represented by an object in the Java or Kotlin programming language, with defined data fields. Rules in Corda are defined in "contracts" attached to specific states, and a state contract is only applied to transactions that contain that state in its inputs or outputs. This mainly concerns the unusual visibility model of Corda, in which transactions can only be seen by their counterparts or those whose subsequent transactions they affect.

3.2.4.2.2. Contracts and messages

The second approach, which will be called the contract-message model, is used by Hyperledger Fabric and Ethereum. Here, multiple "smart contracts" or "chaincodes" can be created on the blockchain, each of which has its own database and associated code. The contract database can only be altered by its code, rather than directly by blockchain transactions. This design pattern is similar to the code and data encapsulation technique in object-oriented programming.

With this model, a blockchain transaction starts with a message sent to a contract, including a few optional parameters or data. As a response to the message and parameters, the contract code is executed and the code is allowed to freely read and write its own database in reaction. Contracts can also send messages to other contracts, but cannot directly access third-party databases. In the terminology used for relational databases, contracts act as applied "stored procedures", where all access to the database involves a predefined code.

Fabric and Quorum, a variant of Ethereum, further complicate this analogy by allowing networks to define various *channels* or *private states*. The main point is to mitigate problems with blockchain privacy by creating separate environments, each of which is only visible to a particular subset of participants. While this appears to be promising in theory, in reality, contracts and data of each private channel or state are isolated from those of the others. Therefore, in terms of smart contracts, these environments are equivalent to separate blockchains.

3.2.4.2.3. Example of rules

Let us examine how transaction rules should be implemented with these two models for a single-asset financial ledger. Each row in our ledger database has two columns containing the owner's address and the amount of assets owned. In the input–output model, transactions must satisfy two conditions:

– The total amount of assets in the transaction outputs must match the total in its inputs. This prevents users from arbitrarily creating or deleting money.

– Each transaction must be signed by each input owner. This prevents users from spending other people's money without permission.

Together, these two conditions are sufficient to create a simple but utterly feasible financial system.

In the message-contract model, the asset contract includes a mechanism for forwarding a "send payment" message, which accepts three parameters: the sender's address, the recipient's address and the quantity to be sent. The contract replies by executing the following four steps:

– It verifies that the transaction has been signed by the sender.

– It verifies that the sender has adequate funds.

– It subtracts the requested quantity from the sender row.

– It adds this quantity to the recipient row.

If one of the checks in the first two steps fails, the contract will be canceled and no payment will be made.

Thereby, the input–output and contract-message models prove to be effective ways for defining transaction rules and protecting a shared database. Indeed, on a theoretical level, any of these models can be utilized to simulate the other. In practice, however, the most appropriate model will depend on the application under construction. To what degree is the data affected by each transaction, is it a limited or significant amount of information? Should we be able to guarantee transaction independence? Is there a clear owner for every piece of data, or is there an overall state that should be shared?

Exploring how the answers should guide our choice between these two models is not our place here. As a general rule however, when developing a new blockchain application, it would be advisable to try to express its transaction rules based on both models and conclude which one fits more naturally. The difference will be reflected in terms of storage and throughput requirements, programming ease and how fast conflicts are detected.

3.2.4.2.4. Built-in rules

Concerning transaction rules, MultiChain differs in one specific way from Fabric, Ethereum and Corda (Table 3.3). Unlike these other platforms, several abstractions are integrated into MultiChain that provide the basic building blocks for blockchain-based applications, without requiring developers to write their own code. These abstractions cover three usually necessary areas: dynamic permissions, transferable assets and data storage.

For example, MultiChain manages the permissions for connecting to the network, sending and receiving transactions, creating assets or streams, or controlling other users' permissions. Multiple fungible assets can be securely and atomically issued, transferred, withdrawn or traded. Any number of "streams" can be created on a chain to publish, index and recover on-chain or off-chain data in JSON, text or binary format. All transaction rules for these abstractions are available and ready to use.

When developing a MultiChain application, this built-in functionality can be ignored and transaction rules can be expressed using smart filters only. However, smart filters are designed to operate with their built-in abstractions, allowing their default behavior to be restricted by way of customized choices. For example, permission for some activities can be controlled by specific administrators, rather than using the default behavior that any administrator would be able to do. The transfer of certain assets may be time-limited or even require additional approval above a certain amount. The data of a particular stream can be validated to ensure that they consist only of JSON structures containing required fields and values.

In all of these cases, smart filters create additional requirements for transaction validation, but do not cancel the simple built-in rules. This can help address one of the main challenges of blockchain applications: the fact that an existing bug in *on-chain* code streams can lead to serious consequences. We have seen countless examples of this problem in Ethereum's public blockchain, and in particular with the DAO hack and parity multisignature bugs. Broader investigations have found a large number of common vulnerabilities in Ethereum smart contracts that enable attackers to steal or freeze other people's funds.

MultiChain smart filters can of course also contain bugs, but their consequences are more limited. For example, built-in asset rules prevent a user from spending someone else's money or accidentally making their own money vanish, regardless of further logic contained in a smart filter. If a bug is found in a smart filter, it can be disabled and replaced with a fixed version, while the basic integrity of the ledger is protected. In philosophical terms, MultiChain is closer to traditional database architectures where the database platform provides a number of built-in abstractions such as columns, tables, indexes and constraints. More powerful features such as triggers and stored procedures can optionally be coded by application developers if actually necessary.

Transaction rules	Fabric	MultiChain	Ethereum	Corda
Model	Contract-Message	Input-Output	Contract-Message	Input-Output
Built-in	None	Permissions, assets, streams	None	None

Table 3.3. *Comparison between Hyperledger, MultiChain Ethereum and Corda transaction rules*

3.2.4.3. *Determinism*

We shall now start the next section of our discussion. Regardless of the chosen approach, custom transaction rules in a blockchain application are expressed in computer code form, written by application developers. Unlike centralized applications, this code is going to be executed more than once and in various locations for each transaction. The reason for this is that several blockchain nodes belonging to different participants must each verify and/or run this transaction themselves.

This repeated, redundant code execution introduces a new requirement seldom found in centralized applications: determinism. In the context of computation, determinism means that a piece of code will always give the same answer for the same parameters, independently of where or when it is executed. This is absolutely crucial for the code that interacts with a blockchain because without determinism, consensus between the nodes in that chain can be catastrophically broken.

Let us now examine what really happens in practice, first in the I/O model. If two nodes have a different opinion on the validity of a transaction, one will accept a block containing that transaction and the other will not. Since each block is explicitly linked to a previous block, this will create a permanent "fork" in the network, in which one or more nodes will not accept the opinion of the majority on the entire blockchain content from that point forward. Minority nodes will be cut off from the scalable state of the database, and will no longer be able to use the application efficiently.

Now let us see what happens if consensus fails in the contract-message model. When two nodes have a different opinion on how a contract should respond to a particular message, this may cause the content in their databases to differ. This in turn can affect the contract response to future messages, including messages it sends to other contracts. The end result is an increasing discrepancy in the way the various nodes view the database state. The "state root" field in Ethereum blocks ensures that any difference in contract responses immediately leads to an absolutely catastrophic blockchain fork, rather than run the risk of masking for some time.

3.2.4.3.1. Sources of non-determinism

Non-determinism in blockchain code is therefore clearly a problem. However, if the basic computational blocks, such as arithmetic, are deterministic, what should we be afraid of? It turns out that there is a variety of possible sources:

– random number generators, since by definition these are designed to produce a different result each run;

– the current time, because nodes will not process transactions at exactly the same time, and in any case, their clocks may be out of sync (time-dependent rules can always be implemented by referring to timestamps in the blockchain itself);

– the use of external resources such as the Internet, disk files or other programs running on a computer. There is no guarantee that these resources will always give the same response and they might possibly become unavailable;

– running several code blocks in parallel "threads", as this creates "race conditions" where the order these processes terminate cannot always be anticipated;

– floating point computations that can provide different answers, even very small ones, on different processor architectures.

Our four blockchain platforms use several different approaches to avoid these drawbacks (Table 3.4).

3.2.4.3.2. Deterministic execution

We shall start with Ethereum, since its approach is the "purest". Ethereum contracts are expressed in a special format called *Ethereum bytecode*, which is executed by the Ethereum EVM virtual machine. This code is compiled from a programming language derived from JavaScript called Solidity. Other languages were previously available but have since become obsolete. Determinism is guaranteed by the fact that Solidity and Ethereum do not support any non-deterministic operations.

MultiChain filters and Corda contracts follow a different approach by adapting existing programming languages and execution environments. MultiChain uses JavaScript running in Google's V8 engine, which is also at the core of the Chrome browser and the Node.js platform, but where non-determinism features are disabled. Similarly, Corda utilizes Java or Kotlin, both of which are compiled into *JAVA bytecode* that runs on a Java virtual machine (JVM). For now, Corda uses Oracle's standard non-deterministic JVM, but it is undergoing development to incorporate a deterministic version. Meanwhile, Corda contract developers should ensure not to allow non-determinism in their code.

The main advantage of Ethereum is risk minimization – a virtual machine built for this purpose is less likely to contain an unintended source of non-determinism. Although software updates would correct such inattention, it would cause problems to any channel that encountered it. Ethereum's problem, however, is that Solidity and the EVM are a tiny ecosystem that take its roots in the broader context of programming languages and runtime environments. In comparison, JavaScript and Java are the two main languages on Github, running on billions of computer devices, the running times of which have been optimized for decades. This is likely why the Ethereum public blockchain is considering a transition to eWASM, a deterministic fork of the emerging WebAssembly standard.

3.2.4.3.3. Determinism by endorsement

To address determinism, Hyperledger Fabric follows a completely different approach. In Fabric, when a client node wishes to send a message to blockchain code, it first sends that message to a group of specific nodes called *endorsers*. Each of these nodes independently executes the blockchain code, thus forming an opinion about the impact that the message will have on the database of that blockchain code. These judgments are sent back to the client along with a digital signature, which amounts to a formal amendment. If the client receives enough mentions of the expected outcome, it creates a transaction containing those mentions and broadcasts it to be included into the chain.

In order to ensure determinism, each piece of chaincode has an *approval policy* that exactly defines the approval level required to make its transactions valid. For example, a blockchain code policy may state that at least half of the nodes comprising the blockchain should give their approval. Another may require the approval of at least one-third of the trusted parties. In any case, each node can independently verify whether the necessary approvals have been received.

Just to make the difference clearer, determinism in most blockchain platforms is based on the question: What is the result of running this code with these data? And we have to be absolutely sure that every node will give the same answer to this question. In contrast, determinism in Fabric hinges on a different topic: Do enough endorsers agree on the result obtained by running this code with these data? Answering this question is a fairly simple matter of counting, and there is no room for non-determinism.

Fabric's approval-based determinism leads to a few interesting consequences. First, blockchain code can be written in a large variety of programming languages because these do not have to be specifically adapted to determinism (Go, Java and JavaScript are currently supported). Second, blockchain code can be concealed to some of the blockchain participants, because it should only be executed by clients and endorsers. The database itself is globally visible. Finally, and most importantly, Fabric chaincode can do things that are not allowed in other blockchain environments such as checking the weather using an online web API. In the worst-case scenario, where each endorser gets a different response from this API, the client will fail to obtain enough endorsements for a particular outcome and no transaction will take place. It should be noted that the Fabric team encourage the use of deterministic logic within the blockchain code to avoid any unpleasant surprises.

What is the price that Fabric pays for this flexibility? If the purpose of a blockchain is to avoid intermediaries for an application based on a shared database, then Fabric's reliance on endorsers is a step away from that goal. For blockchain participants, this means that it is no longer a question of simply following the rules of the blockchain code. It is also necessary that other nodes certify that they are complying with the rules. A malicious subset of endorsers could approve changes to the database that do not follow the blockchain code at all. This gives endorsers much more power than conventional blockchain validators that can suppress transactions but cannot break blockchain rules. Blockchain application developers must decide if this trade-off makes sense in their particular case.

3.2.4.4. *Conflict prevention*

So far, we have explained how different blockchain platforms express transaction rules in code and how they deterministically ensure that each node equally applies those rules. A third aspect shall now be discussed: How does each platform address a situation in which two transactions, both valid on their own, conflict with one another? In the simplest example, assume that Alice has €10 in a financial ledger and broadcasts two transactions: one sending €8 to Bob and the other sending €7 to Charlie. Obviously, only one of these transaction will be allowed to succeed.

Determinism	Fabric	MultiChain	Ethereum	Corda
Model	Amendments	Adapted running time	Dedicated virtual machine	Adapted running time
Languages	Go, Java, JavaScript	JavaScript	Solidity	Java, Kotlin
Code visibility	Counterparts, endorsers	Blockchain	Blockchain	Counterparts + dependent
Forced	No	Yes	Yes	No (for now)

Table 3.4. *Comparison of the deterministic approach in Hyperledger Fabric, MultiChain, Ethereum and Corda*

3.2.4.4.1. Two models

We can start by considering MultiChain's and Corda's approach to this problem altogether. As discussed earlier, both of these solutions use an I/O model to represent transactions and their rules, in which each transaction input uses a previous transaction output. This infers a simple principle for avoiding conflicts: each product can only be spent once. MultiChain filters and Corda contracts can rely on their respective platforms to strictly apply this restriction. Since Alice's €10 are represented by a previous transaction output, this one-time spending rule automatically prevents her from sending it to Bob and Charlie.

Despite this shared feature, the key difference in the way MultiChain and Corda prevent conflicts needs to be stressed. In MultiChain, each node sees every transaction and can therefore independently verify that each output is spent only once. Any double-spending transaction relative to a previously confirmed transaction will be instantaneously and automatically rejected. In contrast, Corda does not implement any global blockchain, thereby *notaries* are required to avoid this double spending. Each output state in Corda is assigned to a notary that must sign any transaction spending that output, thus corroborating that it was not previously spent. Blockchain participants have to trust notaries to honestly follow this rule and malicious notaries can willingly cause damage. As with Fabric amendments, this *one-time spending as a service* has benefits in privacy terms but reintroduces intermediaries, which counterbalances any gain obtained with blockchain. It is important to explain that Corda notaries can be managed by groups of participants based on a consensus algorithm so that the integrity of the chain can be protected from individual bad stakeholders.

Ethereum relies on contracts and messages rather than inputs and outputs. As a result, transaction conflicts such as Alice's two payments are not immediately visible to the blockchain engine. Instead, they are detected and blocked by the contract that processes transactions, after their on-chain command is confirmed. When processing each of Alice's payments, the contract checks if her balance is adequate. If the transaction paying Bob €8 arrives first, it will be processed as usual, leaving Alice with €2 in her account. As a result, when the contract processes the second transaction that should pay Charlie €7, it finds that Alice does not have the necessary funds and cancels the transaction.

3.2.4.4.2. Results versus contracts

Until now, we have seen two different techniques to avoid conflicting transactions: single-expense outputs in MultiChain and Corda, and contract-based verification in Ethereum. One question then comes to mind: Which of these two solutions is the best one?

In order to find the answer to this question, consider an example of an account with €100 in it belonging to Gavin and Helen, and where one of the owners is allowed to independently spend that money. Gavin issues a request to pay €80 to Donna, and a few seconds later, Helen wants to send €40 to Edward. Since there are not enough funds for both payments, these transactions would inevitably create a conflict. If both transactions are broadcast, the outcome will be determined by the first one to enter the chain. Unlike Alice's previous example, this conflict is accidental since no one is trying to violate the application rules: this is merely a time-related issue.

Considering the likelihood that this conflict will occur, we may ask ourselves: Once Gavin sends his transaction, how long will it take for Helen's node to know that her payment might fail? If this time is short, it is less likely that Helen will try to make this payment, thus avoiding the unpleasant notification that her transaction was rejected.

With the input–output model, any conflict between transactions can directly be seen on the blockchain platform since both transactions will explicitly attempt to spend the same previous output. In MultiChain, this happens as soon as Gavin's transaction has propagated to Helen's node, usually in less than a second. With Corda, the output notary will refuse Helen's request to sign the transaction since it has already signed Gavin's: as such, Helen will instantly know that her payment will fail. If the Corda notary is itself distributed, it may have to wait a few seconds to get a response. In any case, there is no need to wait for a transaction to be confirmed and propagated through the blockchain.

In Ethereum, there is no immediate way for the blockchain platform to know that a conflict is going to occur. While Helen's node can see Gavin's transaction on the network, it has no means of knowing how it will affect Helen's transaction because, from its perspective, these are simply two messages sent to the same contract. Once the final command of the conflicting transactions is confirmed on the blockchain, Helen's node will recompute the actual result instead of the expected result and its application will accordingly update the outcome. In the meantime, Gavin and Helen will have no idea of what is happening.

However, we should not conclude that the input–output model always works best. Let us consider a different situation with our example scenario, where Gavin and Helen both require, at exactly the same time, payments smaller than €40 with an initial balance of €100.

In the I/O model, these transactions would conflict, as they both spend the same database row containing those €100 and only one payment would succeed. However, in Ethereum, both transactions would be successfully processed, regardless of their final command since the account has enough funds for both. Here in this specific case, Ethereum's outcome is more in line with Gavin and Helen's intentions.

3.2.4.4.3. Read–write sets

Finally, in Fabric, whose endorsement-based approach is a hybrid of these two techniques: when a Fabric client node wants to send a message to a contract, it first seeks approval from certain nodes to forward that message on its behalf. Approval nodes do this the same way as Ethereum, that is, by executing the contract on their local database, but this process is observed rather than immediately applied. Each endorser records the set of rows that would be read and written, also recording the exact version of those rows at that exact moment. This *read–write set* of versioned lines is explicitly referenced in the endorsement and included in the transaction broadcast by the client.

Conflicts between fabric transactions are solved after their command is completed in the chain. Each node independently processes each transaction, verifying approval policies and applying specified database changes. Nonetheless, if a transaction reads or writes a version of a database row that has already been modified by a previous transaction, the second transaction is ignored. Back to Alice's conflicting payments to Bob and Charlie, these two transactions will read and alter the same row version, containing the €10 Alice started with. Therefore, the second transaction will be automatically and safely abandoned.

Fabric's approach to conflict resolution works seamlessly, but in terms of performance and flexibility, it combines the worst of the previous two models. Given that endorsements convert transactions into specific read–write sets, Gavin and Helen's simultaneous but compatible €40 payments would result in a conflict that Ethereum prevents.

However, Fabric lacks the advantages of the I/O model in speed, because endorsers execute contracts with the most recent version of the blockchain-confirmed database, ignoring unconfirmed transactions. Therefore, if Helen initiates her payment seconds after Gavin, but before it is confirmed on the blockchain, Fabric will create conflicting transactions that a pure I/O model is able to avoid. A comparison summary of these different approaches is proposed in Table 3.5.

Conflict prevention	Fabric	MultiChain	Ethereum	Corda
Model	Sets of read/write	Single expense	Contract controls	Single expense
Verification	Independent	Independent	Independent	Trust notaries
Performance (speed)	~10 s (confirmation)	~1 s (propagation)	~10 s (confirmation)	0–5 s (notary)

Table 3.5. *Conflict prevention management in Hyperledger Fabric, MultiChain, Ethereum and Corda*

3.2.4.5. *Summary*

We have examined several different ways MultiChain, Ethereum, Fabric and Corda address the main challenges involved with *smart contracts*. Each platform provides different answers to our main three questions:

- How are transaction rules represented?
- How is code deterministically executed?
- How can conflicts be prevented?

Each platform represents a complex multi-directional trade-off between flexibility, simplicity, performance, disintermediation, security and privacy. Therefore, the choice of platform for a specific application should first begin with a detailed understanding of the trust model of this application, the types of transactions it involves and their likely conflict patterns.

3.3. State of blockchain research for Industry 4.0

Traceability does not merely involve recording data. Other aspects are involved in traceability such as confidentiality guaranties and storage efficiency, as well as transparency concerning other stakeholders in the supply chain. In addition to security issues (Mullet et al. 2021), several articles in the literature have investigated how blockchain technology is used to address these questions (Chen et al. 2022; Shah et al. 2022). Blockchain is defined as a distributed ledger of digital events or transactions achieved between several parties, and which can be verified at any time in the future. For example, Galvez et al. (2018) presents a list of blockchain solutions to traceability problems, mainly related to the coordination of individual activities, transaction decentralized validation, legitimacy and preservation. By providing transparency (Chauhan et al. 2022) and immutability, the blockchain paradigm successfully contributes to creating a unique level of credibility for all stakeholders and helps build stronger bonds with customers as well as attracting new ones.

Security and safety are other challenges mentioned, but not actually addressed in this research. In order to provide in-depth insights into this technology, Bodkhe et al.

(2020) proposes a review of the implementation of blockchain-based solutions for a variety of applications in which security is of prime importance. He describes how blockchain technology could solve the issues faced by traditional systems in terms of transparency, centralization, scalability, trust and security. The way in which a blockchain can be used to support *smart manufacturing* is also discussed in Mohamed and Al-Jaroodi (2019). The authors propose a middleware for the development of secure, trustworthy, traceable and reliable applications.

The blockchain-secured intelligent manufacturing system presented by Leng et al. (2021) employs blockchain to solve some common problems in manufacturing systems, such as operation traceability, privacy and trust. Moreover, it also makes it possible to avoid faults in key nodes that could occur in centralized platforms. This system is based on the international ISA95 standard (www.isa.org), frequently referenced in Industry 4.0 in the field of information technology. The document also presents some measures related to the use of blockchain in industrial manufacturing, namely concerning data provenance transparency, system flexibility, system sustainability and cost savings (Ko et al. 2018). Privacy is only mentioned as a future blockchain research avenue.

In order to investigate the accuracy of blockchain technology for real-time transparency and cost savings in manufacturing, Ko et al. (2018) provide a comparison of the benefits achieved by two manufacturing companies. Several aspects are considered in this study, including blockchain deployment costs (Pincheira Caro et al. 2022) which are often ignored, and some other limitations related to this technology. These issues of implementation costs and profits are also examined by De Giovanni (2020) in their supply chain game model comprising two companies, namely a supplier and a retailer, using blockchain and *smart contracts*. Business risks, transaction costs and stochastic cases in which deploying blockchain is not advantageous are some of the aspects studied. Another proposal from Lee et al. (2019) consisted of a three-tier blockchain architecture for cyber-physical systems (CPSs) for addressing the challenges associated with the implemention of the 5C-CPS structure. In terms of security and storage, the architecture is an attempt to ensure integrity, fault tolerance, resilience, confidentiality and transparency.

With the advent of Industry 4.0, new manufacturing paradigms have emerged, such as cloud manufacturing. However, some problems related to centralized industrial networks still need to be solved. Consequently, a blockchain-based peer-to-peer (P2P) network architecture was presented by Li et al. (2018). The main goals of this architecture were to make data sharing secure and improve cloud manufacturing reliability and flexibility. Ensuring data reliability is a key obstacle that prevents taking advantage of big data-related analytics tools for improving supply chain efficiency. Thus, Fernández-Caramés et al. (2019) underline that it is essential to determine whether the information collected from the sensors is valid or not. To solve this problem, a combination of UAVs (unmanned aerial vehicles) and a blockchain-based system is presented for the administration of inventory and traceability applications in big data-based supply chain management. However, while blockchain is regarded as a promising technology, the public disclosure of any transaction is considered a security risk by most organizations (Syed et al. 2022).

A blockchain-based privacy preserving system for transaction processing systems is presented in Wang and Kogan (2018) and provides some solutions to these problems. The main idea is to achieve a trade-off between the advantages of sharing information (transparency) and costs associated with weakened privacy. Maintaining privacy in the blockchain refers to the fact that only authorized parties can access sensitive or proprietary transaction data. The proposed system is based on homomorphic encryption and a recent innovation called *zero-knowledge proof*. We share the same goals, while avoiding zero-knowledge proof which introduces higher processing costs and may not be the most adequate in practice for a wide range of data types utilized by manufacturing processes.

3.4. Our approach to blockchain for traceability in Factory 4.0

This section presents our approach to integrating the use of blockchain into the traceability framework. This approach is called the Blockchain based Privacy-preserving and product-Centered Approach for Traceability (BPCAT) and will be referred to throughout this chapter by the acronym BPCAT.

3.4.1. *Target objectives and guarantees*

Several works on the application of blockchain to traceability were already mentioned in section 3.3. Blockchain technology is best known for its immutability and transparency, making it an ideal solution for addressing traceability challenges in Industry 4.0 (De Giovanni 2020). Yet, these features alone do not suffice to address all traceability concerns. Instead, in our work, immutability will be used as a central feature around which other concepts will gravitate to increase their reliability. Our blockchain approach aims at fostering trust between participating stakeholders and providing important benefits for product traceability, namely concerning the following aspects.

3.4.1.1. *Security*

In manufacturing, traceability data are associated with privacy concerns, it is therefore important to enforce this privacy in our approach. In addition, a factory must have full control over the validation process of the blockchain participants that access its traceability system. Unlike conventional public blockchain solutions in which participants can virtually remain anonymous, a private blockchain such as MultiChain, used in this work, only allows a small number of users to access and use the blockchain. And finally, the criticality of some data may impose restrictions on sites where they can be stored and the conditions under which they should remain accessible.

3.4.1.2. *Transparency*

In a blockchain, each validated transaction generates a transaction hash that can be accessed by any participant. It is thus possible to verify that a specific action has been carried out in the blockchain at a given time.

3.4.1.3. *Integrity*

Integrity designates the maintenance of data consistency, accuracy and reliability throughout its lifecycle. To ensure integrity, a hash is produced for each traceability data item and stored in the blockchain. Since blockchain is tamper-proof, the hash can be considered tamper-proof by extension. Integrity verification consists of calculating the data hash and comparing it with the original hash stored in the blockchain.

3.4.1.4. *Confidentiality*

Transparency does not mean that all information must be publicly available. Indeed, traceability data may include some private information related to the manufacturing process. Consequently, we consider that confidential data must be appropriately managed; for example, through encryption before insertion into the blockchain or by inserting in them special data derived from confidential data instead of the confidential data itself. However, the data hash inserted into the blockchain regarding confidential data must be computed from the original data before encryption or any other operation. The reason is that authenticity and integrity will have to be verified against the actual data. In this way, it will be possible to ensure both confidentiality (by means of data encryption) and transparency (by means of derived data and hashing the original data). When derived data are preferred over encryption, the protocol producing that derived data must be exposed in a fully transparent manner and must guarantee that the original data cannot be retrieved from the derived data. The protocols offered in zero-knowledge proof approaches provide such guarantees. However, these are complex approaches and may not be practical in the context of manufacturing plant traceability due to the amount of data and the fact that they are produced in real-time during product manufacturing and often at a very fast pace. Newer approaches to homomorphic encryption could be the ideal solution in the near future.

3.4.1.5. *Non-repudiation*

It ensures that a participant cannot deny any of their actions, whether data-related or blockchain transactions. Since participants are authenticated in a private blockchain, specific mechanisms must be introduced to verify and keep track of data and transaction validation. Digital signatures could be one of the mechanisms possibly used. These signatures make it possible to sign data with a private key and the signature can be verified by any blockchain member using the corresponding public key.

Figure 3.2 represents an application of the proposed product-centric traceability approach. Two traceability steps are represented.

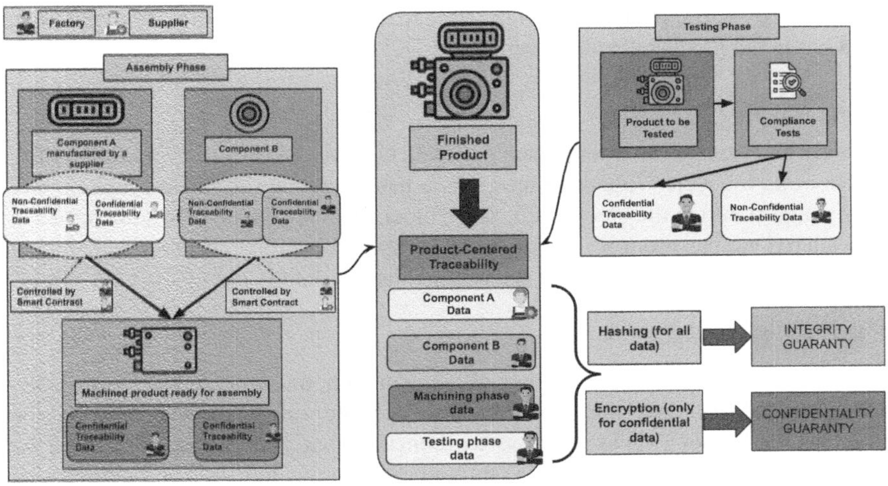

Figure 3.2. *Conceptual overview of the product-centered traceability approach*

During the assembly phase, two components are added to a product that has already undergone several transformations, where component A was manufactured by a supplier while component B was manufactured inside the factory. During the testing stage, compliance tests are performed by the factory and generate confidential test data. In the middle of the figure can be seen the finalized product that aggregates all the traceability data with different confidentiality levels (confidential or not) and different stakeholders (supplier or factory). All these data will serve to produce a corresponding hash in order to guarantee its integrity later in the blockchain. However, only confidential data will be encrypted (encryption or derived).

3.4.2. *Architecture overview*

BPCAT is an architecture that integrates two aspects: product traceability and integration into the blockchain. This overall architecture is represented in Figure 3.3, which will be described from top to bottom.

Three blockchain nodes represent the types of stakeholders or participants, namely the validator (the company that owns the traceability data, in this case the factory), a customer (retailer, end-user, etc.) and a service provider (component supplier, service provider, etc.) involved in product development. These nodes together form the blockchain network, which can be remotely monitored by a node called a manager node. The purpose of the manager is to give more control to the organization that owns the factory (or to an operator designated by the participants) in the blockchain, in particular to manage memberships, authorizations and possibly parameters.

This blockchain implementation is therefore similar to a consortium blockchain, more suitable for the context of the supply chain; this means that all of the identities of its stakeholders are known and the blockchain is controlled not by a single entity but by a consortium of stakeholders that can participate in the blockchain in different ways through the permissions system. The nodes are also deployed for each of the stakeholders (factory, suppliers and clients). This ensures increased transparency, trust between stakeholders and system availability for each party. Conversely, a public blockchain would have been a fully transparent working tool, but would have posed legal, ethical and confidentiality problems, while a private blockchain, implying unilateral control by only one of the participants, could have posed transparency and availability problems for third parties.

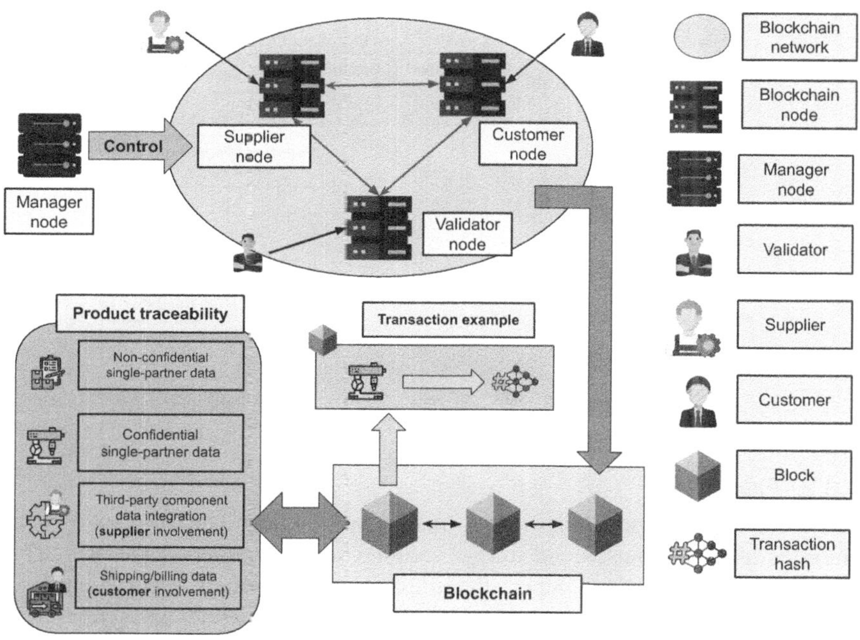

Figure 3.3. *BPCAT general architecture*

3.4.3. *Data management*

3.4.3.1. *Raw data*

In the blockchain, every time new traceability data is added to a block, a transaction is created. BPCAT imposes certain prerequisites on how the data should be structured, as shown in Figure 3.4. A typical blockchain transaction always consists of a transaction hash and a timestamp to ensure that an action was carried out at a specific time in the blockchain. On the other hand, the transaction hash does not guarantee data integrity. For this reason, BPCAT also appends the data hash to the transaction. With the data hash, the timestamp and the transaction hash, transaction transparency and data integrity can be guaranteed to third parties (factory, suppliers, clients).

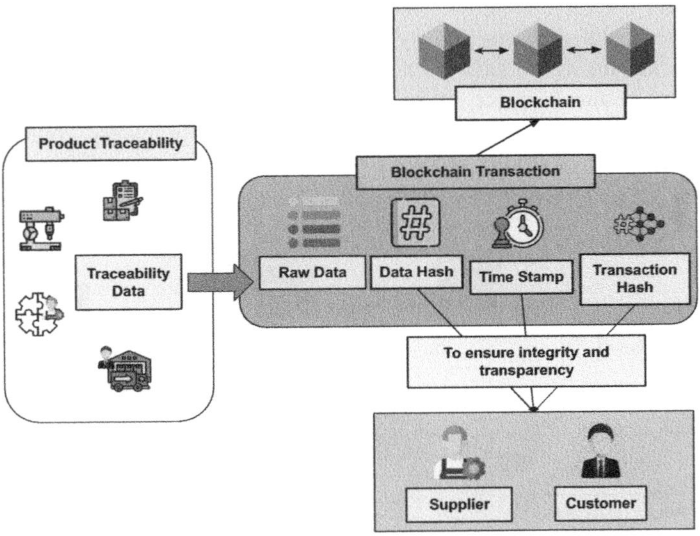

Figure 3.4. *BPCAT raw data management*

3.4.3.2. *Files*

In order to deal with the problem of high traceability data volumes that may require the use of an outsourced storage solution such as the cloud, the BPCAT file management policy does not impose that files be stored inside the blockchain.

Each participant can use an external storage solution as can be seen in Figure 3.5. However, the file hash and file path are stored in the blockchain; the first is used to ensure file integrity and authenticity, and the other is to easily retrieve the file. This way, any of the members can resort to cloud storage if needed.

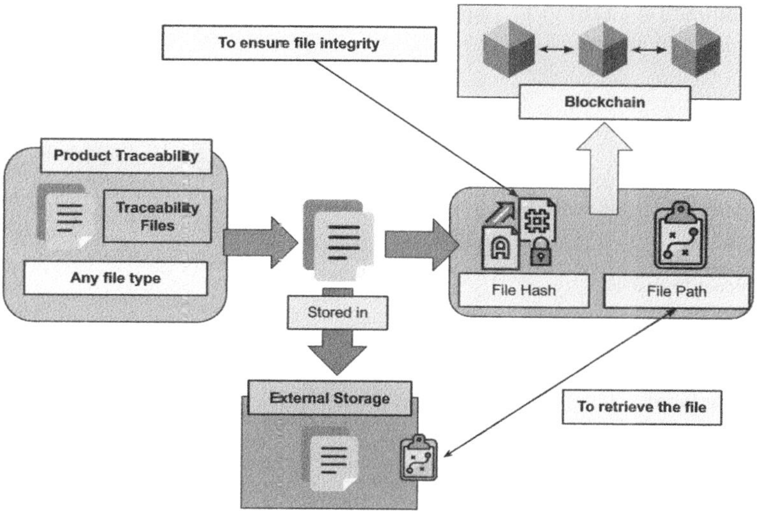

Figure 3.5. *BPCAT file management*

3.4.4. *Privacy management*

Confidential traceability data refer to data that the factory owner (or another participant vis-à-vis the other stakeholders) does not wish to reveal to all the blockchain participants, but which must be included in the traceability data as they might be needed as evidence during retrospective investigations.

Indeed, in order to avoid conflict with the transparency usually expected from a blockchain solution, BPCAT handles confidential data in the following way:

– Confidential data are encrypted using an encryption method chosen by the data owner to ensure confidentiality. This will typically be an encryption according to the latest recommendations in force and the cryptographic capabilities of the systems producing the traceability data.

– In order to implement transparency, BPCAT suggests that any encrypted data in a transaction should include related information that would facilitate its authenticity and integrity to be evaluated if necessary. The hash of the original data before encryption, or the hash of the data derived from the original data in a predefined way, can be used for this purpose. All these data are inserted into the transaction, and the transaction timestamp and hash serve as assurance of the authenticity and integrity of the entire transaction.

– In the event of a dispute, a participant may request the data be decrypted by a trusted agent imposed by the procedures in place or accepted by the author of the data to compare its hash to the hash stored with the data in the blockchain, and thus establish its authenticity.

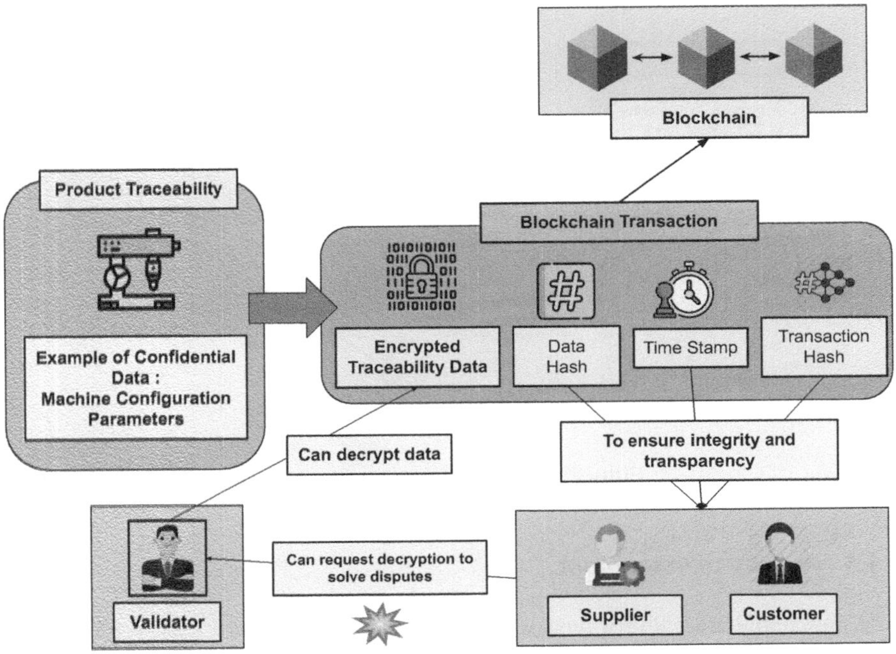

Figure 3.6. *Privacy management with BPCAT*

Figure 3.6 shows that encrypted data can only be decrypted by the owner (in this case the validator) in order to maintain confidentiality. However, both the supplier and the client have access to the original data hash, to the timestamp and the transaction hash, which guarantees integrity, authenticity and transparency. The time duration that data are exposed as well as the frequency with which the encryption keys are changed should be set according to the factorization risks in the duration or according to the volumes generated with the same keys.

3.4.5. *Non-repudiation management*

Regarding traceability, it can be useful for resolving disputes where the responsibility of one of the players is questioned. For this purpose, digital signatures can be used in the blockchain, as shown in Figure 3.7. For each blockchain transaction involving traceability data, each stakeholder (validator, client, supplier) involved in the lifecycle of the associated product must sign the transaction. Using their personal private key, each intervening party signs a message containing the transaction hash and a timestamp to authenticate the signature. Then, once the digital signature is computed, it is registered in the blockchain. Non-repudiation is guaranteed through all the signatures, which can be verified by anyone using the signer's public key.

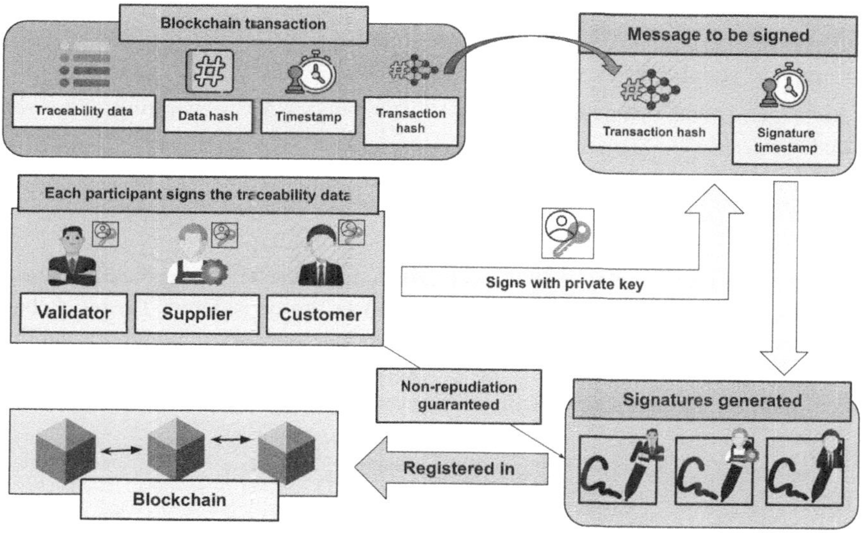

Figure 3.7. *Non-repudiation management with BPCAT*

3.5. Functional description of the BPCAT MultiChain implementation

In order to put our approach into practice, a prototype was developed using the MultiChain[1] tool. It is an *open source* platform designed to create permissioned blockchains. The main advantages of this platform include the control over all blockchain aspects (permission management, consensus algorithms, data storage, etc.), the possibility to create multiple blockchains and its ease of use for developers.

We will focus here on the prototype features that directly implement the approach; for completeness, a more technical description is available in section 3.7.

1 See: https://www.multichain.com.

3.5.1. *Network architecture and stakeholder representation*

In MultiChain, the blockchain participants are called "nodes", and the connections between these nodes form the blockchain network. With BPCAT, nodes can be compared to stakeholder avatars in the traceability system. The proposed implementation comprises a total of five nodes:

– validator: factory avatar, traceability system owner;

– customer: customer avatar;

– supplier: supplier avatar;

– manager: node managing the state of other nodes and generating addresses/private keys;

– archival: node storing old blockchains and allowing them to be a posteriori consulted, in particular for audit purposes.

Figure 3.8 is a representation of the network architecture including the interactions between the blockchain participants. Three networks are considered: the blockchain network formed by the blockchain nodes, the corporate network in which traceability data is generated and submitted to the blockchain, and the external network as an abstraction of all the interactions with entities external to the factory.

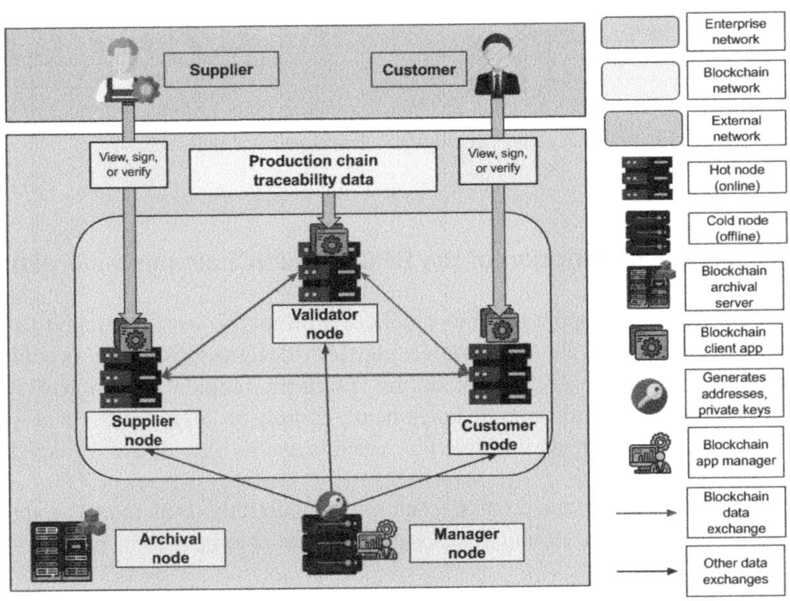

Figure 3.8. *Implementation overview of the network and nodes*

The hot nodes (validator, customer, supplier) form the blockchain network and are connected to each other. Operating as the primary node, the validator will represent the manufacturing plant and publish the traceability data. The customer and supplier nodes will act as the entry points for transparency. As such, they will allow customers and suppliers to access traceability data.

In contrast with the previous nodes, the manager node represents a "cold" MultiChain node: this is an isolated node that is not connected to the blockchain. From the blockchain standpoint, this node does not actually exist. Its function is to monitor the blockchain status and of other nodes. The reason why it was initially designed is to address one of the main problems often mentioned by companies concerning the blockchain, essentially: the blockchain cannot be easily managed due to its decentralized nature. Remote control consists of starting and stopping other nodes, or more generally instantiating a new blockchain. Moreover, it is also responsible for generating the addresses of other nodes, as well as encryption, decryption and data signing private/public keys. An external public key infrastructure (PKI) collectively or contractually chosen by the partners would be able to implement some of these functionalities.

The function of the archive node is to store old blockchains and it allows them to be utilized at a later point. Figure 3.9 shows the three blockchain nodes and how they are monitored from the manager node in MultiChain.

Figure 3.9. *Node status visualization from the manager node*

Finally, it should be noted that due to the specific composition of the blockchain nodes including the factory and its partners, the consensus can be simplified in order to avoid unnecessary energy consumption.

3.5.2. *Blockchain and storage management*

As mentioned earlier, MultiChain implements a simple mechanism for creating and initiating blockchains. In BPCAT, this feature is used to avoid another drawback of blockchain technology for Industry 4.0 factories, namely its characteristic "ever-growing" nature. We suggest a time criterion defining the blockchain lifespan before shutting it down and initiating a new one. That way the blockchain size is stable, and the data volume as well as the storage cost will be easily estimated and anticipated. Then, the retention of the history of older chains is managed by the archiving node, and data storage includes deleting blockchains whose age exceeds a value defined according to the factory traceability policy.

3.5.3. *Raw traceability data management*

In order to manage and store data, MultiChain implements a blockchain concept abstraction for data storage and management called *streams*. These are data containers operating as key-value storages. Each piece of data is called an item, and an item is a key-value pair in which each key is used to index and retrieve a specific item. An item can have multiple keys, and the same key can be used multiple times. Figure 3.10 provides an overview of how traceability data are organized in the implementation as well as the different data types encountered.

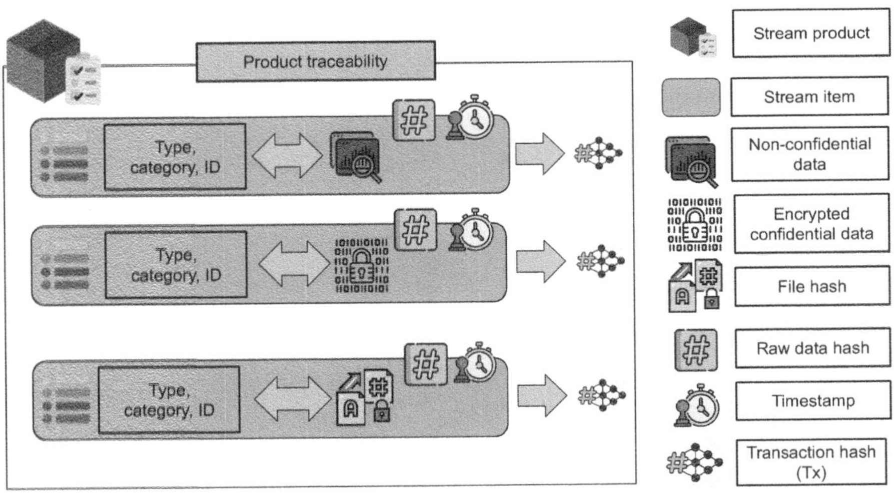

Figure 3.10. *Raw data management implementation in BPCAT*

In the approach implementation, we use one stream per product in which the stream name can include the product serial number or any other unique product attribute. Each piece of product traceability data is represented in the form of an item whose keys are its type or category. For each item, the data hash is stored so that the integrity and authenticity of the data can be verified.

In the snippet of Listing 3.2, the structure of an item can be seen as it is submitted to MultiChain by the JSON prototype. This item consists of two parts: a header containing the data type, timestamp, encryption level and hash named *data_hash*. On the other hand, the body corresponds to the traceability data.

```
{
    "header": {
        "data_type": "TRACEA_STEP",
        "encrypted": false,
        "timestamp": 1660060004,
        "data_hash":
            "eaf19d9b4ad0cdcf125c395e1c292df9e3a7cc5c5fc3b529
            97270ba64bb7e515"
    },
    "body": {
        "location": "Line9-u-3",
        "location_entry": 1660063000.73701,
        "location_exit": 1660064354.73701,
        "process_desc": "Reduced discrete Internet solution"
    }
}
```

Listing 3.2. *Example of raw traceability data in a MultiChain stream*

3.5.4. *Privacy guaranty*

In our approach, we consider that the data can be divided into two categories in security terms: confidential and non-confidential data. To manage this difference, confidential data must be encrypted before it is inserted into the blockchain. Figure 3.11 describes how data privacy is protected in the MultiChain implementation. Hybrid encryption based on symmetric encryption (AES 128) and asymmetric encryption (RSA 2048) is used. First, the data are encrypted using a unique encryption-generated random symmetric encryption key. This encryption key is then encrypted with asymmetric encryption to produce what is known as an encrypted encryption key (CC key). The CC key can only be decrypted by those who possess the associated private key. The CC key is stored in a dedicated stream as Tx hash storage–CC key, where Tx Hash is the transaction hash produced when the encrypted data was stored in the blockchain. Thereby, transaction data decryption consists of retrieving the CC key linked to the transaction hash, decrypting it with the asymmetric private key, and decrypting the data with the final decrypted key thus obtained.

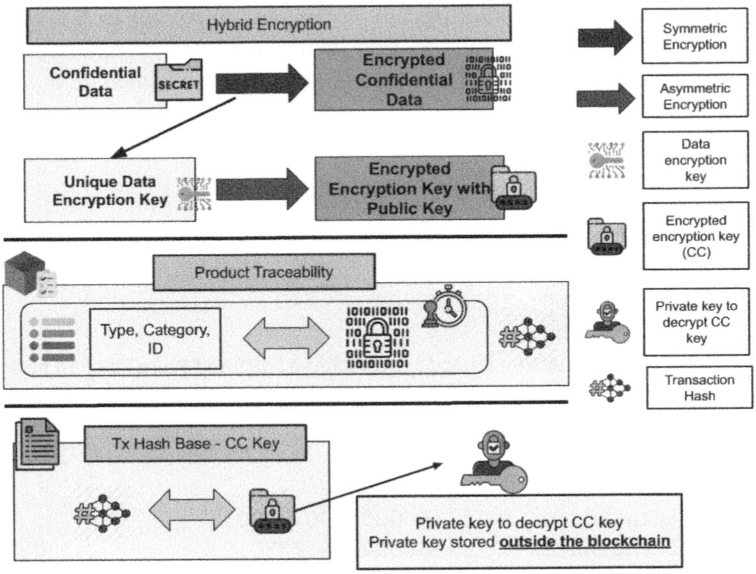

Figure 3.11. *Confidentiality implementation in BPCAT*

The snippet in Listing 3.3 presents the *encrypt_bytes* function, which illustrates this approach. The function can be used to encrypt a data array in byte form; as a result, it returns the encrypted data *ink_data* as well as the CC key represented by the variable *ink_aes*, returned as a *StreamItemSecret* object to be inserted into a MultiChain stream.

```
def encrypt_bytes(self, databytes):
    aes_key = generate_aes_key(AES_KEY_SIZE)
    rsa_pub = read_rsa_key("pub")
    encr_aes = encrypt_rsa_msg(rsa_pub, aes_key)
    encr_data, tag, nonce = encrypt_aes_msg(aes_key,
    ↪ databytes)
    item_secret = StreamItemSecret(
        secret=bytes_to_str(encr_aes, encode=True),
        tag=bytes_to_str(tag, encode=True),
        nonce=bytes_to_str(nonce, encode=True),
    )
    return encr_data, item_secret
```

Listing 3.3. *Date encryption function in the BPCAT implementation*

3.5.5. *File management*

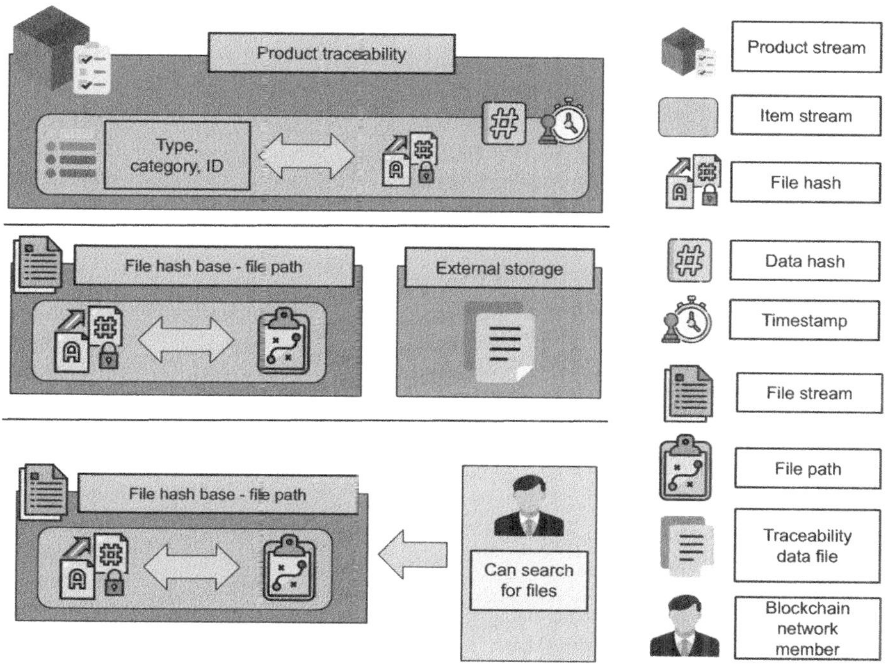

Figure 3.12. *File management implementation in BPCAT*

Product traceability data can take many forms, namely that of a file. In our approach, files are not directly stored in the blockchain. In fact, the company is free to choose the storage method that best meets its needs.

Figure 3.12 describes how traceability files are managed and how they are interconnected with the traceability data described in the previous section. In the MultiChain implementation, file information is stored in a dedicated stream. Items are coded as a file hash–file path pair and for each item, a transaction hash is produced. Finally, to reference the file in product traceability, the transaction hash mentioned above is used.

The snippet of Listing 3.4 shows the *save_file_bytes* function as an illustration of this file management. The function computes the file hash, then writes the file to a given directory that can represent external storage. Finally, file information such as its type, size, path is registered into the blockchain in the form of a transaction.

```
def save_file_bytes(self, filepath, filename, save_dir,
        filetype):
        filehash = hash_file(filebytes)
        filepath = os.path.join(save_dir, filename)
        write_bytes(filepath, filebytes)
        return self.save_file_infos(
            StreamFile(
                filetype=filetype,
                filename=filename,
                filehash=filehash,
                filesize=len(filebytes),
                offchain=False,
            )
        )
```

Listing 3.4. *Function for saving a file with MultiChain*

3.5.6. *Data signing*

MultiChain makes it possible for a blockchain participant to sign data and allows for verifying the validity of a signature. In our approach, the signature reflects the fact that a data item has been approved by a different blockchain participant than the one that published it. Figure 3.13 shows how signatures are stored in MultiChain and how their security and reliability are ensured.

Signatures are stored in a dedicated stream where each element is a transaction–signature hash pair where the transaction hash corresponds to the hash of the transaction to be signed. In order to secure the process and guarantee the uniqueness of all signatures, signature generation is based on multiple parameters such as the timestamp, the address of the node performing the signature and the transaction hash. These three pieces of information therefore enable us to answer the questions “Who? When? What?” Figure 3.14 illustrates how data are signed in the BPCAT implementation with MultiChain.

Furthermore, to ensure that only valid signatures are added to the blockchain, a stream filter enables signature verification before adding it (Listing 3.5). This helps maintain the reliability of both the main blockchain and the blockchain client application.

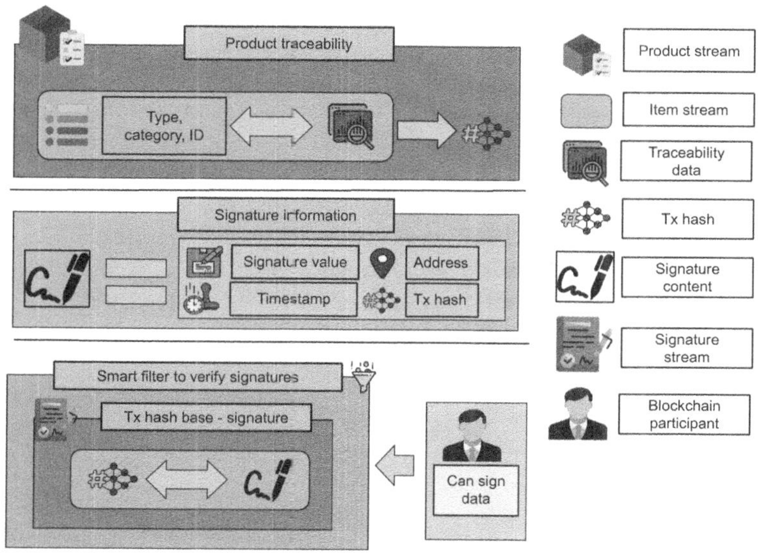

Figure 3.13. *Signature implementation in BPCAT*

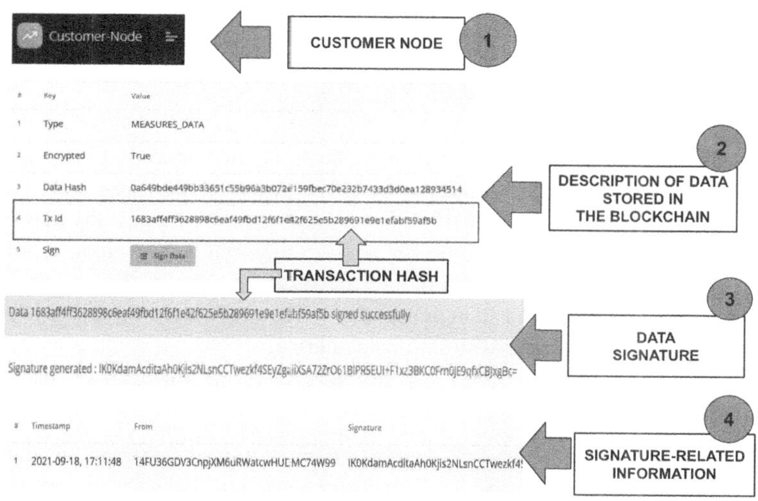

Figure 3.14. *Data signing example in BPCAT*

```
const SIGNATURES_STREAM = "SIGNATURES";

function filterstreamitem() {
  stream = getfilterstream();
  streamitem = getfilterstreamitem();

  if (stream.name == SIGNATURES_STREAM) {
    address = streamitem.data.json.body.from_addr;
    txid = streamitem.data.json.body.for_tx;
    signature = streamitem.data.json.body.signature;

    if (!verifymessage(address, signature, txid)) {
      return (
        "Signature " + signature +
        " is invalid from address " + address +
        " for tx " + txid
      );
    }
  }
}
```

Listing 3.5. *Smart filter for signature authenticity verification*

3.5.7. *Data integrity verification*

For all traceability data, the raw data hash is also stored. Consequently, it is possible to guarantee data integrity from the moment they are added into the blockchain and every time they are used. In our approach, we enable any participant to verify data authenticity. For example, for non-confidential textual data, it is possible to copy and paste the data and recompute the hash directly in the prototype client interface. If the computed hash matches the one stored, data integrity is verified. Regarding confidential data, two situations can arise:

– If the requester is able to decrypt the data, nothing differs from the previous example.

– However, if the data cannot be decrypted, the requester must send a "decryption request" to the data owner, most likely the company that owns the traceability data. Then, if the request is approved, the decrypted data should simply be copied and pasted and the computed hash compared with the stored hash.

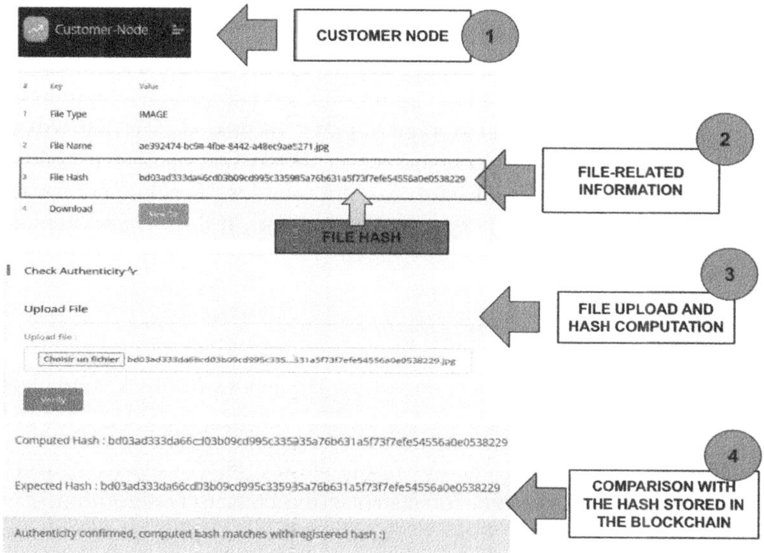

Figure 3.15. *Data integrity verification example with BPCAT*

The file integrity verification process (Figure 3.15) is quite similar. It requires downloading the file to be verified in order to compute its hash. This hash is then compared to the stored hash. If the two values match, file integrity is ensured.

Listing 3.6 shows a snippet of the function for verifying if two files are similar and thus for confirming integrity.

```
def check_file_integrity(filetxid):
    stream_file = services.file.get_file_by_txid(filetxid)
    form = CheckFileHashForm()
    file = form.file.data
    expected_hash = stream_file.data.body.filehash
    computed_hash = hash_file(file.read())
    if computed_hash == expected_hash:
        flash_success("Files are similar")
    else:
        flash_error("Files are not similar")
```

Listing 3.6. *Snippet of the function for file integrity verification*

3.6. Technical challenges to be addressed and proposed solutions

In addition to its advantages, blockchain technology presents several technical challenges, the main ones being block mining-related energy consumption, reducing chain-level as well as node-level storage volume. In this section, we will propose several solutions to overcome them based on different scenarios tested with the prototype developed with MultiChain. To this end, a "benchmark" system was integrated in order to be able to carry out measurements. It is described in more detail in section 3.8.

3.6.1. *Mining and energy consumption*

Reducing energy consumption is one of the challenges of blockchain technology when dealing with big data.

The main element related to energy consumption is the consensus algorithm used in the mining process. In Bitcoin, for instance, the consensus algorithm is proof of work (POW). MultiChain is a "fork" of the Bitcoin core and is therefore capable of using POW as well as the round-robin validation scheme. Round-robin is the default scheme in MultiChain and is based on mining diversity where nodes with mining permission will add blocks, either randomly or one after the other depending on the diversity parameter. There is therefore no competition in this particular consensus algorithm. Regarding proof of work, nodes are made to compete with each other in order to solve a mathematical puzzle as quickly as possible, which explains the high energy consumption. To illustrate how the consensus algorithm impacts power consumption, a comparison of the CPU usage rate with MultiChain when using POW and round-robin is presented in Figure 3.16.

For each node, two curves are available: POW and non-POW (round-robin). Next to the non-POW curves the gains can be seen with respect to the POW curve associated with each node. For example, using a round-robin scheme results in a –59% difference for the validator, –83% for the supplier and –82% for the customer. A discussion can therefore be initiated about energy consumption. Therefore, if a factory considers that a blockchain is necessary to its traceability policy and justifies the additional energy consumption associated, some of its suppliers and customers may not agree with its position. Currently, BPCAT employs the round-robin based scheme in order to save energy for each participating player; however, it could also be decided that only the factory (validator node) should perform mining operations to spare suppliers and clients any additional energy costs due to mining.

3.6.2. *Storage volume optimization*

Storage volume is another challenge related to blockchain technology. Given that blockchain is a distributed system, this implies that all participating nodes must store a copy of the blockchain thus resulting in a less efficient solution than a centralized system in which data would be stored only once (in the factory infrastructure, for example). The following sections describe a way of solving this problem on the basis of two parameters: size reduction of the data itself and the reduction of the volume stored on each node.

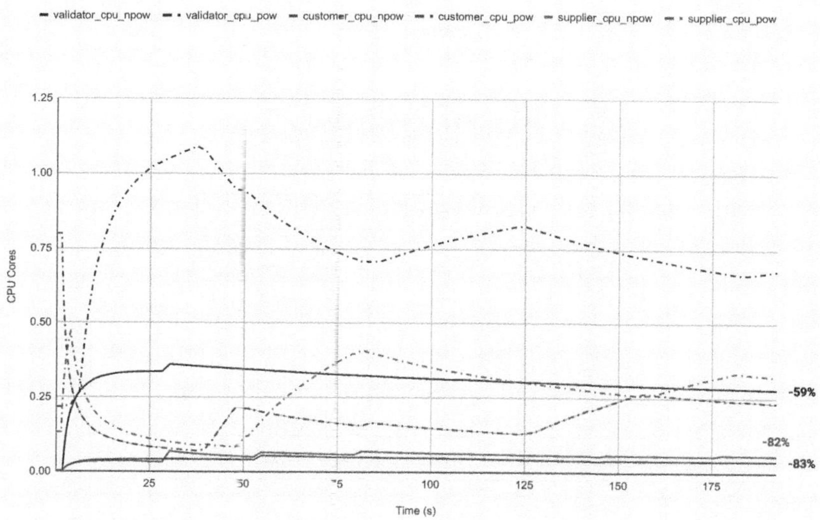

Figure 3.16. *Comparison of CPU consumption between two mining algorithms: proof of work (POW) and round-robin (non-POW)*

3.6.2.1. *Data-scale optimization*

MultiChain provides a mechanism for storing data as *off-chain* items, which means that data are stored outside of the blocks and referenced by a hash, unlike *on-chain* storage in which data are directly stored in the blocks. This is not a neutral choice as can be seen in Figure 3.18.

Storing multiple items per transaction leads to reducing the amount of storage needed, while *on-chain* storage requires almost double the volume compared to off-chain storage. When data are submitted off-chain, they are internally split into fixed-size chunks, and for each chunk, a hash is calculated. This process is similar to how IPFS works (Hasan et al. 2020), which is a distributed file storage system often associated with blockchain systems for its file management capability. Nonetheless, unlike IPFS, MultiChain natively includes this feature.

The most interesting characteristic of this process is that if there is already a chunk with the same hash, it is not stored a second time and instead a reference to the existing chunk is created. Consequently, storage space can be greatly reduced for data with a high similarity rate since duplicate data are not stored more than once.

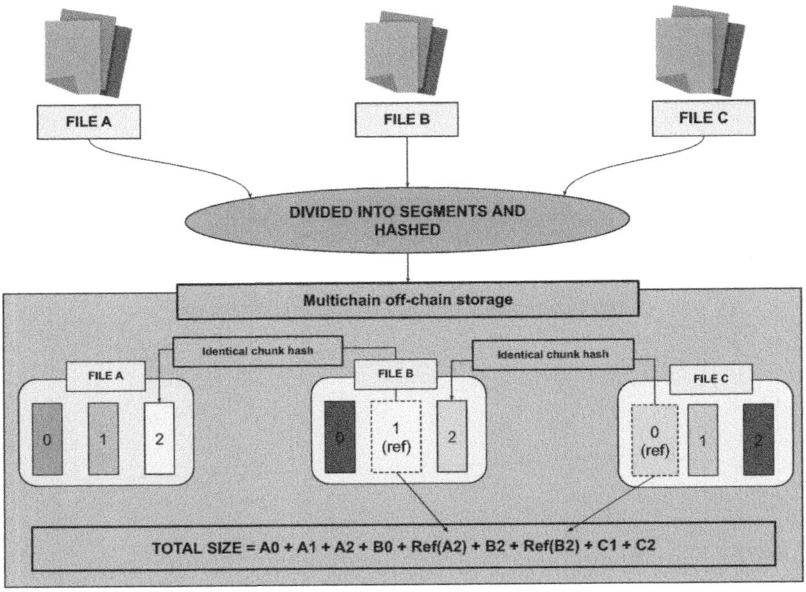

Figure 3.17. *Description of the MultiChain off-chain storage mode operation*

To evaluate to what extent this feature can allow storage savings, we conducted an experiment with a large volume of files exhibiting different similarity rates and then submitted them to MultiChain in off-chain mode.

The context is as follows: a thousand images with a 1 MB size are generated with various difference rates ranging from 5% to 100%. Listing 3.7 shows the function that, using an original pixel array, generates another array other than N%, where N is the variance rate in this function.

The result of this experiment is illustrated in Figure 3.19 which shows the storage volume optimization based on the variance rate between files.

The curve labeled "Total initial volume" indicates the total storage volume required for the files simply be stored on a hard drive. The other curves represent seven situations where the variance between files varies from 5% to 100%. A 5% variance means that the 1,000 files differ from each other by only 5%, and a 100% variance means that the files are completely different from each other. The storage volume savings in each situation relative to the initial total volume are shown as a percentage reduction on the corresponding curve. This reduction ranges from –90% when the files differ by only 5% to +1% when the files are completely different (a 100% deviation). In the last example, the volume is greater than the initial total volume due to the additional information introduced by MultiChain in file storage management.

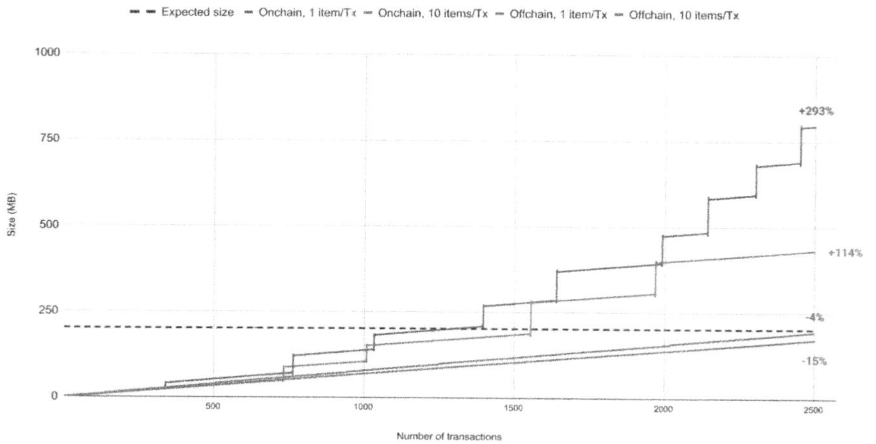

Figure 3.18. *Storage volume optimization in MultiChain according to storage mode (on-chain/off-chain) and the number of items per transaction*

```
def get_alternate_pixels(orig_pixels, variant_rate=5):
    alternate = orig_pixels.copy()
    nb_update = round(len(alternate) * variant_rate / 100)
    for idx in range(nb_update):
        rand_pixel = services.random.rand_int(min_value=0,
            max_value=255)
        alternate[idx] = rand_pixel
    return alternate
```

Listing 3.7. *Function for creating an image with a N% difference from an original image*

The manufacturing plant framework implies that the same processes are indefinitely repeated. Consequently, the likelihood of finding similarities between the data generated by product-centered traceability processes is very high. Therefore, the feature for splitting data into chunks and not saving the same chunk twice could result in reducing the total volume required to store traceability data.

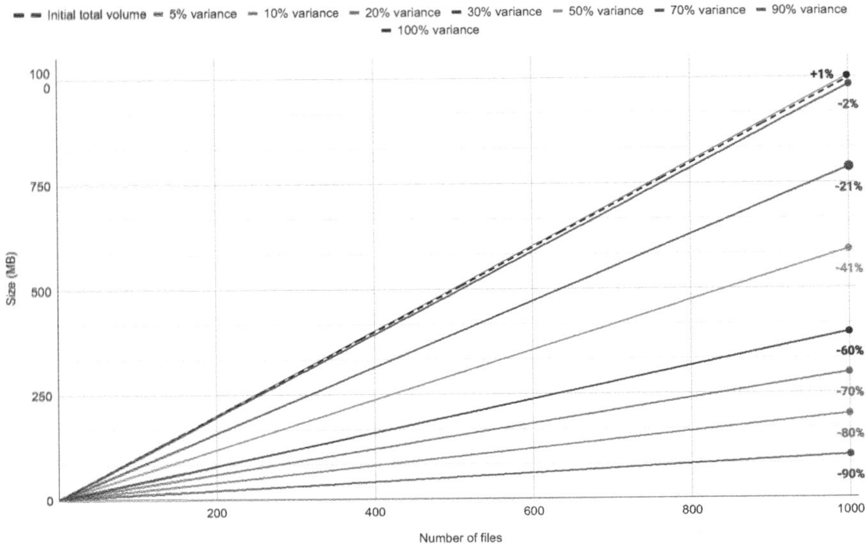

Figure 3.19. *Storage volume optimization in MultiChain based on the variance rate between files (off-chain mode)*

3.6.2.2. *Node-level optimization*

The second problem is the volume of data stored in each node. The idea that the blockchain is a distributed ledger might suggest that each node should store exactly the same amount of data in its blockchain copy, which is not entirely necessary. As mentioned earlier, MultiChain can record data in streams, thus allowing the blockchain to be used as a general-purpose database providing only timestamping, notarization and immutability. By default, each node must store transaction blocks and hashes. However, data stored in streams requires specific storage management due to the subscription feature. Subscribing to a stream involves indexing and uploading all of the content of the stream to the node, including off-chain data. This means that each stakeholder can be somewhat flexible when choosing what data should be stored in local storage, which can lead to saving a significant volume of data.

To illustrate the extent of this flexibility, we performed an experiment to demonstrate the possibility of optimizing the storage volume in the BPCAT MultiChain implementation from the stream subscription.

The context can be described as follows: three specific streams were created for each player, and each subscribed to its own stream (validator, customer and supplier). A fourth node called the manager (observer) was created, without subscribing to any stream in order to show the minimal storage volume needed by any node participating in the blockchain network.

Note that 10,000 100 kB file transactions were randomly sent to these streams as shown in Listing 3.8.

```
def start_file_simu_for_stream():
    # List of possible streams
    for_streams = ["VALIDATOR", "CUSTOMER", "SUPPLIER"]
    folderpath = os.path.join(DATAFILES_DIR, filetype)
    filelist = os.listdir(folderpath)
    for i in range(nbfiles):
        filepath = os.path.join(folderpath, filelist[i])
        # Random stream selection
        for_stream = for_streams[
            round(random.randint(0, 100) * (len(for_streams)
               - 1) / 100)
        ]
        bench.measure_perf(
            func=services.publish.prepare_file_request,
            filepath=filepath,
            encrypted=encrypted,
            publish_now=True,
            for_stream=for_stream,
        )
```

Listing 3.8. *Snippet from the function for submitting a transaction to a random stream*

In Figure 3.20, summarizing the results, the first curve shows that the total volume stored by the validator node is the highest, which is explained by the fact that it is the one that submitted all the transactions (the transaction publishers have no choice but to store their own data). However, the customer and the supplier, respectively, observe a –52% and –73% volume difference compared to the validator. The observer node has the lowest volume (–99.96% in comparison to the validator's storage volume).

It therefore appears that by clearly defining the relationships between data and stakeholders, they can be efficiently organized in the streams in order to save a large volume of storage on each node.

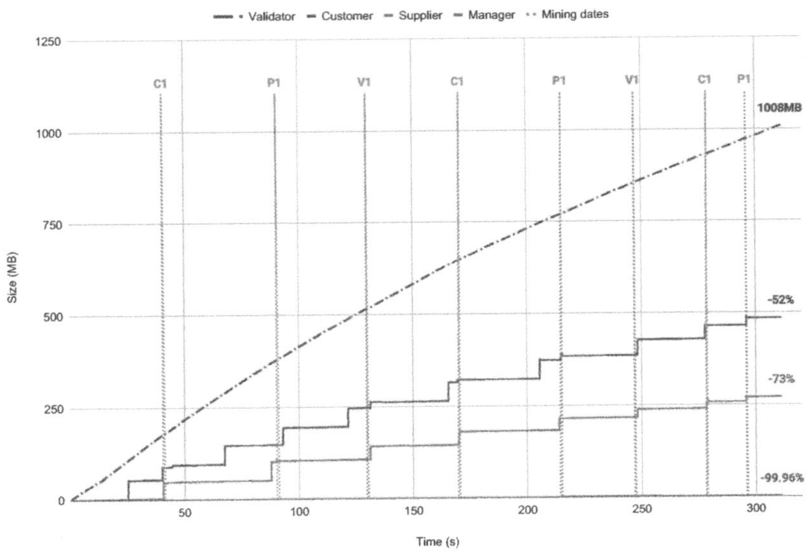

Figure 3.20. *Node-level storage volume optimization according to MultiChain stream subscription*

3.7. BPCAT MultiChain implementation technical description

The various results presented in this chapter concerning the MultiChain BPCAT prototype, such as mining-related measurements, required the development of related tools, whose functionalities will be described in this section. It is therefore possible for the reader to directly use our work as a basis to create a BPCAT implementation adapted to the reader's own needs, and to reproduce our experiments as well as our results for multiple purposes.

The first difficulty posed by the blockchain architecture representation lies in its distributed nature. The concern is that the development of a blockchain prototype involves using a technology that would make it easy to deploy several nodes, knowing that each node represents a MultiChain instance. Therefore, we should have a tool that would be able to quickly deploy several instances with different configurations both at the application level and the network level. These capabilities are precisely integrated in the Docker technology; this is thus our choice as the basic technology for developing the prototype.

Figure 3.21 illustrates this architecture where it can be seen that each node is represented by two Docker containers: a container containing a MultiChain instance and a container with the client application for the interaction with the MultiChain API. This client application was developed in *Python*.

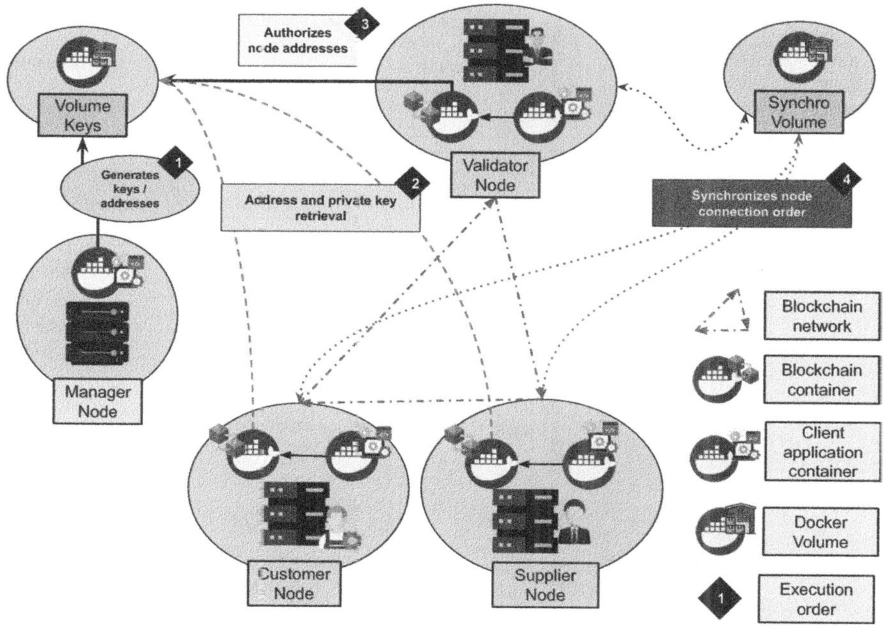

Figure 3.21. *BPCAT MultiChain prototype software architecture*

Each node has a well-defined role in the architecture and follows a specific hierarchy; this is why they need to be able to exchange data, especially during the stage before the blockchain initialization. With that in mind, Docker volumes were used, since they can be simultaneously mounted on several containers thus allowing information to be shared. For example, the *Key Volume* serves to store the private and public keys generated by the manager node.

Since this is a distributed network and the blockchain can only be initialized by the validator, it obviously appears that a synchronization and an initiation command between nodes must be followed: this role is performed by the *Synchro Volume*.

3.7.1. *Node configuration and deployment*

3.7.1.1. *Configuration*

With Docker, a configuration can be separately set up for each container through environment variables. Table 3.6 illustrates the list of environment variables used in the validator blockchain container.

Name	Description	Example
NODE_ROLE	Name of the role that the node performs. Multiple nodes can have the same role	validator
NODE_ID	Unique node identifier	validator1
MULTICHAIN_DATADIR	MultiChain directory location	/multichain
KEYPAIRS_DIR	*Key Volume* location	/tmp/keypairs
SYNCHRO_DIR	*Synchro Volume* location	/tmp/synchro
CHAINNAME	Blockchain name	chain1
NETWORK_PORT	Network port used for inter-node communications	7000
RPC_USER	Login to connect to the MultiChain API in RPC	rpcuser
RPC_PASSWORD	Password to connect to the MultiChain API in RPC	secretpassword
RPC_PORT	Listener port used by the MultiChain API	8000
RPC_ALLOW_IP	List of IP addresses allowed to connect to the API (default localhost only)	localhost
CUSTOMER_CAN_MINE	Boolean to define whether or not the right to mine should be granted to the client node	false|true
SUPPLIER_CAN_MINE	Boolean to define whether or not the right to mine should be granted to the supplier node	false|true
WAIT_FOR_NODE	Defines the node to wait for before initiation (to order the boot sequence)	manager-node
PARAM_MINING_DIVERSITY	Used to define the value of the mining-diversity parameter in MultiChain	mining-diversity|0.5
PARAM_TARGET_BLOCK_TIME	Used to define the value of the target-block-time parameter in MultiChain	target-block-time|30
PARAM_MAXIMUM_CHUNK_SIZE	Used to define the value of the maximum-chunk-size parameter in MultiChain	maximum-chunk-size|1024
PARAM_TARGET_ADJUST_FREQ	Used to define the value of the target-adjust-freq parameter in MultiChain (to change the mining algorithm)	target-adjust-freq|-1

Table 3.6. *Description of the blockchain container environment variables*

The customer and supplier nodes only have an additional variable called MASTER_NODE, which is the IP address of the validator node, since they need the validator node to connect to the blockchain.

3.7.1.2. *Deployment*

In order to deploy all the node-related configurations, docker-compose was used; this is a tool associated with Docker for easily deploying multicontainer application configurations, which is the case with the blockchain architecture. The docker-compose configuration consists of providing a file in which the parameters of all the containers to be deployed are indicated.

```
validator1-node-chain:
    build: ./node-validator/chain
    container_name: chain-${VALIDATOR_NODE_ID}
    networks:
      - chain_net
    ports:
      - ${VALIDATOR_PORT_RPC_HOST}:${GL_CHAIN_RPC_PORT}
    volumes:
      - ${GL_VOLUME_SYNCHRO}:${GL_DIR_SYNCHRO}
      - ${GL_VOLUME_KEYPAIRS}:${GL_DIR_KEYPAIRS}:ro
      - ${GL_VOLUME_MULTICHAIN}:${GL_DIR_MULTICHAIN}
    expose:
      - ${GL_CHAIN_NETWORK_PORT}
      - ${GL_CHAIN_RPC_PORT}
    environment:
      NODE_ROLE: ${VALIDATOR_NODE_ROLE}
      KEYPAIRS_DIR: ${GL_DIR_KEYPAIRS}
      CHAINNAME: ${GL_CHAIN_NAME}
      MULTICHAIN_DATADIR: ${VALIDATOR_MULTICHAIN_DIR}
      NETWORK_PORT: ${GL_CHAIN_NETWORK_PORT}
      PARAM_TARGET_BLOCK_TIME: ${GL_CHAINP_TARGET_BLOCK_TIME}
```

Listing 3.9. *Snippet of the validator node docker-compose configuration*

Listing 3.9 shows a snippet of the configuration used for example to deploy the validator blockchain container, which includes several sections:

– *networks*: network to which the node belongs (all nodes must belong to the same network);

– *ports*: list of ports to be redirected (between host and container for example);

– *volumes*: container-level volume to be mounted such as *Key* or *Synchro* volumes;

– *expose*: section containing the list of ports to be opened at container level (blockchain port as well as the RPC API port);

– *environment*: section where the list of environment variables to be defined and their values is inserted.

3.7.2. Blockchain initialization

3.7.2.1. *Address, private and public key generation*

During this stage, the manager node prepares the private and public keys for each node allowing MultiChain to generate the addresses and sign the blocks mined by the node considered. For this, the manager node can use a client application also developed in Python to perform this task. Listing 3.10 illustrates the manager node function for generating node addresses. The manager creates three files in the *Key Volume*, namely, the address value but also the public key/private key pair used by MultiChain for signing transactions and blocks.

```
def generate_address(node_id, key_role, chainname):
    chaindir = os.path.join(KEYPAIRS_DIR, chainname)
    services.init_services(read_rpc_conf(COLD_CHAINNAME, True))
    keypair = services.wallet.createkeypairs(1)
    write_key_tofile(node_id, "addr", key_role, keypair[0].
      get("address"), chaindir)
    write_key_tofile(node_id, "priv", key_role, keypair[0].
      get("privkey"), chaindir)
    write_key_tofile(node_id, "pub", key_role, keypair[0].
      get("pubkey"), chaindir)
```

Listing 3.10. *Manager node address generation*

3.7.2.2. *Address and private key retrieval*

Each container then retrieves the private key addressed to it to integrate it into its MultiChain configuration. The container will therefore import the generated file from the *Key Volume* as shown in Listing 3.11.

```
# KEYPARDS_DIR => "Key Volume"
multichain-cli $CHAINNAME -datadir="$MULTICHAIN_DATADIR"
importprivkey $(head -n 1 $KEYPAIRS_DIR/$CHAINNAME/$NODE_ID
.connect.priv)
```

Listing 3.11. *Third-party node-generated private key retrieval*

3.7.2.3. *Node address authorization*

The validator node initializes the blockchain and allows the different nodes to connect with one another. This results in retrieving the node addresses generated in the *Key Volume* as seen in Listing 3.12.

```
multichain-cli $CHAINNAME -datadir="$MULTICHAIN_DATADIR"
grant $(head -n 1 $KEYPAIRS_DIR/$CHAINNAME/customer.connect
.addr) connect
```

Listing 3.12. *Address recovery for the node generated by the validator node*

3.7.2.4. *Node synchronization and initialization command*

The node initialization sequence is as follows: manager, validator, then customer and supplier. This operation is ensured by each node reporting its status through generating a file in the *Synchro Volume.*

Synchronization consists of waiting for the file to be generated by the previous node to signal that it is ready, as shown in Listing 3.13.

```
# Waiting for the node to be ready
while [ ! -f "$SYNCHRO_DIR/$WAIT_FOR_NODE.ready" ];
do sleep 1; done

# Signal "Ready" status
touch "$SYNCHRO_DIR/$NODE_ID.ready"
```

Listing 3.13. *Synchronization operation between nodes*

3.7.3. *Data pre-loading*

The proposal for a blockchain architecture for traceability involves the prior creation of certain blockchain elements such as *smart filters* or streams to store files and signatures. This is why a data pre-loading system has been set up to automatically create these elements. The system was integrated into the client application and runs with a *JSON* configuration file as can be seen in the example of Listing 3.14.

```
{
        "action": "create",
        "category": "stream",
        "data": {
          "name": "SIGNATURES",
          "options": {},
          "details": { "desc": "Stream to store
             signatures" }
        }
}
```

Listing 3.14. *Configuration for the automatic creation of the Signatures stream*

3.8. Measurements, performance and benchmark

As part of the evaluation of the prototype performance and the influence of some parameters on blockchain operation, a tool destined to measure these indicators was implemented in the client application.

The tool will be referred to as *benchmark* throughout this section.

3.8.1. *Measurement retrieval*

For the retrieving of mining-level statistics, the main metric utilized was the CPU rate used. The latter was retrieved with Docker and the *docker-stats* API provided within. However, this API is only available at the host level and cannot be accessed from the container. Yet, the benchmark is started from the container of the client application.

In order for this information to be accessible to containers, it was decided to export the statistics measured by Docker to files stored in a folder mounted on the client application container.

Listing 3.15 illustrates the recovery of CPU consumption by Docker in which filtering was carried out according to the node of interest (here *validator1*) and the statistic looked for (CPU).

The result is then exported to a file.

```
while true; do
    DOCKER_STATS=$(docker stats --no-stream)
    echo "$DOCKER_STATS" | grep "validator1"
    | awk '{ gsub("%","",$3); print $3}'
    | tr -d '\n' > tmp/stats/validator1.cpu
done
```

Listing 3.15. *Recovery of the CPU usage rate measured by Docker*

The previously generated file is stored in a folder called */tmp/stats* mounted on the client application container, as seen in Listing 3.16.

```
validator1-node-client:
    volumes:
        - /tmp/stats:/tmp/stats
```

Listing 3.16. *Folder mounting procedure to access the statistics measured by Docker*

3.8.2. *Client application measurement reading*

The statistics measured by Docker are now stored in files, one merely has to read them when running the benchmark. Listing 3.17 presents a snippet from the *get_node_stats* function for reading the statistics (CPU, memory) in a given node.

```
def get_node_stats(self, node_id):
        cpu = self.get_stat_value(
            node_id=node_id, stat_type="cpu") / 100
        mem = self.get_stat_value(node_id=node_id,
           stat_type="mem")
        return {
            "{0}_cpu".format(node_id): cpu,
            "{0}_mem".format(node_id): mem
        }
```

Listing 3.17. *Docker stats retrieval in the benchmark*

3.8.3. *Benchmark operating principle*

In the context of blockchain, the benchmark will consist of sending a given number of transactions called *number of iterations* and evaluating when each transaction ends the value of certain indicators such as CPU rate, blockchain size, block mining detection and, if this is the case, the identity of the minor node.

3.8.3.1. *Benchmark profile creation*

Various benchmark functions were created in the prototype, in particular to perform measurements according to data type (raw or file) or according to the confidentiality level (data encryption or no encryption). The function presented in Listing 3.18 illustrates a standard benchmark function where it is possible to choose the number of transactions, the individual data size as well as whether or not to use data encryption.

```
def start_data_simu(simuname, samplesize, nbiterations,
        encrypted):
        bench = Benchmark(name=simuname)
        folderpath = os.path.join(DATAFILES_DIR, "json",
        ↪ samplesize)
        filelist = os.listdir(folderpath)
        for i in range(nbiterations):
            filename = filelist[i]
            filepath = os.path.join(folderpath, filename)
            json_obj = json.loads(read_text(filepath))
            # Send the transaction and statistics measurement
            item = bench.measure_perf(
                func=services.data.save_data,
                data_type="MEASURES_DATA",
                keys=["MEASURES", "MEASURES_DATA"],
                obj_data=json_obj,
                encrypted=encrypted,
            )
            bench.show()
        bench.export_csv()
```

Listing 3.18. *Benchmark for submitting raw data to the blockchain*

3.8.3.2. *Command creation*

In order to use the benchmark in CLI (*Command-Line Interface*) mode, a function has to be defined that calls the benchmark function passing a number of parameters such as data size, the number of iterations and whether or not encryption is used. The snippet of Listing 3.19 shows the command associated with the benchmark function outlined in the previous section.

```
@simu_bp.cli.command("data")
@click.option("-s", "--samplesize", type=str,
    default="1kb")
@click.option("-q", "--nbiterations", type=int)
@click.option("-e", "--encrypted", is_flag=True)
def cmd_data_simu(samplesize, nbiterations, encrypted):
    simuname = "data-{0}-{1}-encr{2}".format(samplesize,
        nbiterations, encrypted)
    start_data_simu(
        samplesize=samplesize,
        simuname=simuname,
        nbiterations=nbiterations,
        encrypted=encrypted,
    )
```

Listing 3.19. *CLI mode benchmark execution command*

3.8.3.3. *Benchmark execution*

It is now possible to run the benchmark in CLI mode. Listing 3.20 shows the command to run a benchmark in the following context: 2,000 iterations based on a data sample with an individual size of 1 kB.

```
flask simu data -s 1kb -q 2000
```

Listing 3.20. *Benchmark to run 2,000 iterations based on a 1kb data sample in CLI mode*

3.9. Summary of the strengths and weaknesses of the contribution

In this chapter, we have seen how blockchain technology can be integrated into the traceability of a Factory 4.0. First, a presentation of the blockchain paradigm allowed us to identify some of the main characteristics: a trusted third party is no longer needed to implement transactions, transparency, but above all immutability through the use of a shared ledger; this means that transactions cannot be modified or deleted. In addition, blockchain adds a qualitative aspect to data due to the fact that they are integrated, dated and their availability is guaranteed due to the distributed nature of the architecture. The distributed network is also a means for greatly reducing the risk of data loss or unavailability due to a local failure at some of the nodes.

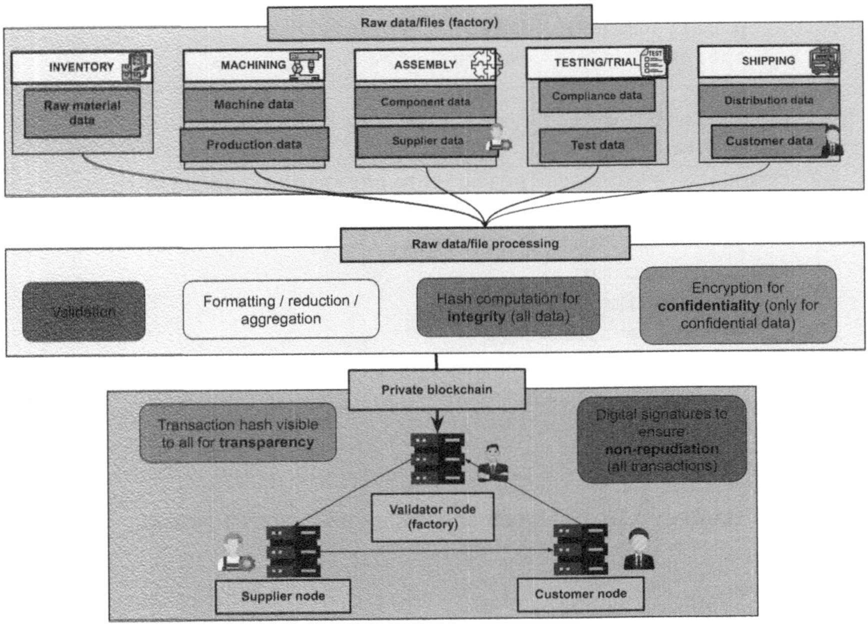

Figure 3.22. *BPCAT approach summary*

In terms of traceability, blockchain can help to securely and transparently share data between every stakeholder in the supply chain. Our BPCAT approach (Figure 3.22), as well as the prototype developed with MultiChain, shows that it is possible to guarantee various critical features related to traceability data, namely:

– security through the implementation of a consortium (or permissioned) blockchain;

– integrity through data hashing and methods for data verification;

– confidentiality through confidential data encryption and the storage of the original data hash;

– transparency by making transactions visible to all stakeholders (but not necessarily the data);

– non-repudiation-based transaction signatures issued by the key players.

All of these elements foster trust and confidence among the stakeholders of the information system, because they are integrated into a shared and tamper-proof transaction ledger known as the blockchain. In addition, the guaranteed features such as the data hash can be retrospectively verified, which can be recomputed to confirm integrity, or transaction signatures, which can be used for authenticity validation. This system can therefore ensure traceability, reduce risks, facilitate management, help solve disputes between stakeholders and improve the sustainability and flexibility of the traceability system in the factory.

Nonetheless, blockchain also presents some frequently mentioned disadvantages. The first relates to energy consumption, which we also discussed and which is largely due to the consensus algorithm employed, with POW being one of the most energy-greedy. We showed that by resorting to other types of consensus this energy impact can be reduced.

The time required for transaction validation is also cited (Caro et al. 2018), particularly in very large-sized public blockchain networks such as Bitcoin. However, our approach allows us to avoid this type of drawback, because we use a permissioned blockchain where the number of participants is limited.

Furthermore, node-level data redundancy can be at the origin of a significant consumption of storage space. This is the reason why we show in the MultiChain implementation that it is possible to minimize this consumption by distributing data across different streams; each player can choose to subscribe to the streams that matter to them so that only the data that are relevant to them are stored locally.

Although blockchain technology is still very young, many of the drawbacks that have been identified are related to errors due to implementation choices (blockchain types, consensus algorithm, configuration of the chosen solution). The development of permissioned blockchains creates new opportunities allowing this technology to be adopted in B2B contexts such as supply chain traceability. This technology is being rapidly adopted in areas where traceability and transparency are legal requirements and demanded by end users, such as in the pharmaceutical (Bocek et al. 2017; Shanley 2017; Leal et al. 2021) and food supply chains (Bumblauskas et al. 2019; Casino et al. 2019; Lin et al. 2019).

Finally, blockchain represents a complete paradigm shift for the industry, which consists of moving from a centralized internal traceability system within the factory to a decentralized network in which relevant players such as customers and suppliers can be integrated. Problems associated with the adoption and integration of this technology into existing ecosystems can obviously emerge. In light of all these factors, means for assessing the impacts of blockchain thus should be made available when integrating it into product traceability for Factory 4.0.

3.10. References

Alkhader, W., Alkaabi, N., Salah, K., Jayaraman, R., Arshad, J., Omar, M. (2020). Blockchain-based traceability and management for additive manufacturing. *IEEE Access*, 8, 188363–188377.

Androulaki, E., Barger, A., Bortnikov, V., Cachin, C., Christidis, K., Caro, A., Enyeart, D., Ferris, C., Laventman, G., Manevich, Y. et al. (2018). Hyperledger fabric: A distributed operating system for permissioned blockchains. In *Proceedings of the Thirteenth EuroSys Conference (EuroSys '18)*. Association for Computing Machinery, New York, 1–15.

Bocek, T., Rodrigues, B., Strasser, T., Stiller, B. (2017). Blockchains everywhere – A use-case of blockchains in the pharma supply-chain. *Computers & Industrial Engineering*, 177, 772–777.

Bodkhe, U., Tanwar, S., Parekh, K., Khanpara, P., Tyagi, S., Kumar, N., Alazab, M. (2020). Blockchain for Industry 4.0: A comprehensive review. *IEEE Access*, 8, 79764–79800.

Bumblauskas, D., Mann, A., Dugan, B., Rittmer, J. (2019). A blockchain use case in food distribution: Do you know where your food has been? *International Journal of Information Management*, 52, 102008.

Buterin, V. (2014). Ethereum whitepaper [Online]. Available at: https://ethereum.org/en/whitepaper/.

Buterin, V. (2015). On public and private blockchains [Online]. Available at: https://blog.ethereum.org/2015/08/07/on-public-and-private-blockchains.

Caro, M.P., Ali, M.S., Vecchio, M., Giaffreda, R. (2018). Blockchain-based traceability in agri-food supply chain management: A practical implementation. In *Proceedings of the 2018 IoT Vertical and Topical Summit on Agriculture - Tuscany (IOT Tuscany)*. IEEE, Tuscany, 1–4.

Casino, F., Kanakaris, V., Dasaklis, T., Moschuris, S., Rachaniotis, N. (2019). Modeling food supply chain traceability based on blockchain technology. *IFAC-PapersOnLine*, 52, 2728–2733.

Castro, M. and Liskov, B. (1999). Practical Byzantine fault tolerance. In *Proceedings of the Third Symposium on Operating Systems Design and Implementation (OSDI'99)*. USENIX Association, 173–186

Chauhan, H., Deepali, G., Gupta, S., Nayak, S., Shankar, A., Singh, P. (2022). Framework for enhancing the traceability in supply chain using blockchain. *Journal of Interconnection Networks*, 22(3), 2144008.

Chen, Y., Lu, Y., Bulysheva, L., Kataev, M.Y. (2022). Applications of blockchain in Industry 4.0: A review. *Information Systems Frontiers*, 26, 1715–1729.

De Giovanni, P. (2020). Blockchain and smart contracts in supply chain management: A game theoretic model. *International Journal of Production Economics*, 228, 107855 [Online]. Available at: https://www.sciencedirect.com/ science/article/pii/S0925527 320302188.

Fernández-Caramés, T.M., Blanco-Novoa, O., Froiz-Míguez, I., Fraga-Lamas, P. (2019). Towards an autonomous Industry 4.0 warehouse: A UAV and blockchain-based system for inventory and traceability applications in big data-driven supply chain management. *Sensors*, 19(10) [Online]. Available at: https://www.mdpi.com/1424-8220/19/10/2394.

Galvez, J.F., Mejuto, J., Simal-Gandara, J. (2018). Future challenges on the use of blockchain for food traceability analysis. *TrAC Trends in Analytical Chemistry*, 107, 222–232 [Online]. Available at: https://www.sciencedirect.com/ science/article/pii/ S0165993618301304.

Hasan, H.R., Salah, K., Jayaraman, R., Ahmad, R.W., Yaqoob, I., Omar, M. (2020). Blockchain-based solution for the traceability of spare parts in manufacturing. *IEEE Access*, 8, 100308–100322.

King, S. and Nadal, S. (2022). Ppcoin: Peer-to-peer crypto-currency with proof-of-stake [Online]. Available at: https://people.cs.georgetown.edu/~clay/classes/fall2017/ 835/papers/ peercoin-paper.pdf.

Ko, T., Lee, J., Ryu, D. (2018). Blockchain technology and manufacturing industry: Real-time transparency and cost savings. *Sustainability*, 10(11) [Online]. Available at: https://www.mdpi.com/2071-1050/10/11/4274.

Leal, F., Chis, A., Caton, S., Gonzalez-Velez, H., García–Gómez, J., Durá, M., Sánchez–García, A., Sáez, C., Karageorgos, A., Gerogiannis, V. et al. (2021). Smart pharmaceutical manufacturing: Ensuring end-to-end traceability and data integrity in medicine production. *Big Data Research*, 24, 100172.

Lee, J., Azamfar, M., Singh, J. (2019). A blockchain enabled cyber-physical system architecture for Industry 4.0 manufacturing systems. *Manufacturing Letters*, 20, 34–39 [Online]. Available at: https://www.sciencedirect.com/science/arti- cle/pii/S2213846319300264.

Leng, J., Ruan, G., Jiang, P., Xu, K., Liu, Q., Zhou, X., Liu, C. (2020). Blockchain-empowered sustainable manufacturing and product lifecycle management in Industry 4.0: A survey. *Renewable and Sustainable Energy Reviews*, 132, 110112 [Online]. Available at: https://www.sciencedirect.com/science/article/ pii/S1364032120304032.

Leng, J., Ye, S., Zhou, M., Zhao, J.L., Liu, Q., Guo, W., Cao, W., Fu, L. (2021). Blockchain-secured smart manufacturing in Industry 4.0: A survey. *IEEE Transactions on Systems, Man, and Cybernetics: Systems*, 51(1), 237–252.

Li, Z., Barenji, A.V., Huang, G.Q. (2018). Toward a blockchain cloud manufacturing system as a peer to peer distributed network platform. *Robotics and Computer-Integrated Manufacturing*, 54, 133–144 [Online]. Available at: https://www.sciencedirect.com/ science/article/pii/S073658451830022X.

Lin, Q., Wang, H., Pei, X., Wang, J. (2019). Food safety traceability system based on blockchain and epcis. *IEEE Access*, 7, 20698–20707.

Masood, F. and Faridi, A.R. (2018). Consensus algorithms in distributed ledger technology for open environment. In *Proceedings of the 2018 4th International Conference on Computing Communication and Automation (ICCCA)*. IEEE, Greater Noida, 1–6.

Mohamed, N. and Al-Jaroodi, J. (2019). Applying blockchain in Industry 4.0 applications. In *Proceedings of the 2019 IEEE 9th Annual Computing and Communication Workshop and Conference (CCWC)*, Las Vegas, NV, 0852–0858.

Mullet, V., Sondi, P., Ramat, E. (2021). A review of cybersecurity guidelines for manufacturing factories in Industry 4.0. *IEEE Access*, 9, 23235–23263.

Nakamoto, S. (2009). Bitcoin: A peer-to-peer electronic cash system. Cryptography Mailing [Online]. Available at: https://metzdowd.com.

Olson, K., Bowman, M., Mitchell, J., Amundson, S., Middleton, D., Montgomery, C. (2018). Sawtooth: An introduction [Online]. Available at: https://www.hyperledger.org/wp-content/uploads/2018/01/HyperledgerSawtoothWhitePaper.pdf.

Pincheira Caro, M.R., Vecchio, M., Giaffreda, R. (2022). Characterization and costs of integrating blockchain and IoT for agri-food traceability systems. *Systems*, 10, 57.

Shah, K., Patel, N., Thakkar, J., Patel, C. (2022). Exploring applications of blockchain technology for Industry 4.0. *Materials Today: Proceedings*, 62, 7238–7242 [Online]. Available at: https://www.sciencedirect.com/science/article/ pii/S2214785322020600.

Shanley, A. (2017). Could blockchain improve pharmaceutical supply chain security? [Online]. Available at: https://www.pharmtech.com/view/could-blockchain-improve-pharmaceutical-supply-chain-security.

Syed, N.F., Shah, S.W., Trujillo-Rasua, R., Doss, R. (2022). Traceability in supply chains: A cyber security analysis. *Computers & Security*, 112, 102536 [Online]. Available at: https://www.sciencedirect.com/science/article/ pii/S0167404821003606.

Szabo, N. (1997). Formalizing and securing relationships on public networks. *First Monday*, 2(9), 1–21.

Wang, Y. and Kogan, A. (2018). Designing confidentiality-preserving blockchain-based transaction processing systems. *International Journal of Accounting Information Systems*, 30, 1–18 [Online]. Available at: https://www.sciencedirect.com/science/article/pii/S146 7089518300794.

4

Impact Assessment of Blockchain on Factory 4.0

Traceability is cross-cutting to the various building blocks of Industry 4.0, as it provides essential information for managing plant operation and facilitates post-production audits. Recently, blockchain technology has been the focus of a large number of studies as a solution for improving the immutability and transparency of industrial traceability systems, with the objective of fostering trust among various stakeholders. However, most studies proposed evaluations that focus on blockchain performance itself rather than its impact on the industrial system. In this chapter, we propose a discrete event-based modeling approach as well as a collaborative framework that can be adapted to any industrial production plant. The objective is to provide manufacturers with means to a priori evaluate through simulation the potential impacts of a blockchain-related project on their production chain, prior to any costly prototyping or deployment.

With the help of a case study, we show how our approach could be used to analyze the impact of our proposal for blockchain-based traceability, namely BPCAT, in terms of different custom metrics, such as storage requirements, energy consumption or even the environmental impact that adopting such a solution would incur.

4.1. State-of-the-art evaluation of factory systems and blockchain

Traceability does not merely involve registering data. Other aspects such as confidentiality and efficient storage are involved, as well as transparency toward the other supply chain players. Aside from security issues (Mullet et al. 2021), several

For a color version of all the figures in this chapter, see: www.iste.co.uk/sondimullet/cybersecurity.zip

Cybersecurity and Traceability in Factory 4.0,
coordinated by Patrick SONDI and Valentin MULLET.

articles in the literature investigated the use of blockchain technology to address these aspects (Chen et al. 2022; Huo et al. 2022). In this section, we focus on the main evaluation methods used in some of them.

A survey of blockchain solutions for sustainable manufacturing and autonomous product lifecycle management is proposed in Leng et al. (2020). Various trust-related measures for the use of blockchain solutions in manufacturing plants are introduced, focusing on transparency, decentralized decision-making, reputation and customer relations. From the plant perspective, blockchain could be used as a means to drive already existing systems for shop floors, such as ERP and MES.

From the product management perspective, blockchain would afford a unified database for sharing product information. Proposals presented in various manufacturing sectors (Tönnissen and Teuteberg 2020), including textiles (Hader et al. 2022), underline the relevance of practical blockchain-based solutions for traceability in Industry 4.0. In order to provide in-depth knowledge about blockchain technology, Bodkhe et al. (2020) present a review of implementations of blockchain-based solutions for various applications in which security is of the utmost importance. They describe how blockchain technology could be used to solve the problems faced by conventional systems in terms of transparency, centralization, scalability, trust and security. The blockchain-secured smart manufacturing system presented by Leng et al. (2021) employs blockchain to solve certain common problems in manufacturing systems, such as operational traceability, privacy and trust. Furthermore, it can be used as a mechanism to prevent key nodes from possibly failing in centralized platforms. The document presents several manufacturing blockchain-related parameters, namely data source transparency, system flexibility, system sustainability and cost savings. However, these inquiries do not provide any quantitative evaluation.

In order to trace the aggregated spare parts in a manufactured product, Hasan et al. (2020) present a blockchain-based approach that addresses some traditional problems met by supply chain management, such as operation transparency and active monitoring that affect the compliance and safety of the components of the spare parts. The proposal involves a file storage method based on the decentralized InterPlanetary File System (IPFS) for spare parts data, smart contract algorithms, security and cost analysis. The evaluation method mainly consists of listing the qualitative advantages of blockchain to ultimately demonstrate the reliability of the spare parts ownership tracking solution. A cost analysis of *smart contract* algorithms is provided based on transaction and execution costs in ether/gas/dollars. This evaluation is however specific to the proposed architecture and cannot be generalized.

Another example of blockchain-based product traceability is proposed by Alkhader et al. (2020) for additive manufacturing. The goal was to ensure secure and reliable traceability, as well as accessibility and immutability. The proposed architecture uses Ethereum and IPFS smart contracts as a distributed file storage solution. The evaluation presented by the authors focuses on security requirements such as integrity, accountability, non-repudiation and clearances. However, it does not include an analysis of the costs related to the use of IPFS as a storage solution.

In order to investigate the accuracy of blockchain technology for real-time transparency and cost savings in the manufacturing industry, Ko et al. (2018) propose a comparison of the gains achieved by two manufacturing companies. In this study, several aspects are taken into consideration, in particular the costs of initializing the blockchain that are often ignored, and several other limitations related to this technology. The proposed assessment is based on a mathematical model that establishes that there are incentives for a manufacturing plant in implementing blockchain technology, without including any case studies to support this conclusion.

Concerns about cost and profit are also examined by De Giovanni (2020) in their supply chain game consisting of two companies, namely a supplier and a retailer, which makes use of blockchain and *smart contracts*. The topics addressed therein include business risks, transaction costs and stochastic cases in which deploying blockchain is not advantageous. However, some industry professionals may find that the evaluation method consisting of game theory equations would be hard to follow and therefore adapt it to their context.

Another proposal from Lee et al. (2019) is a three-tier blockchain architecture for cyber-physical systems (CPSs) to address the challenges associated with the implementation of the 5C-CPS architecture. The architecture aims to achieve integrity, fault tolerance, resilience, confidentiality and transparency, at the security and storage levels. The authors however do not provide any evaluation to support their contribution.

With the advent of Industry 4.0, new manufacturing paradigms such as cloud manufacturing have emerged. However, some problems related to centralized industrial networks still need to be solved. Consequently, a blockchain-based peer-to-peer (P2P) network architecture was presented by Li et al. (2018). The main goals of this architecture were to secure data sharing and improve reliability and flexibility in cloud manufacturing. Ensuring data reliability is a major issue when leveraging big data analytics to ensure supply chain efficiency. The evaluation only takes into account the scalability of the blockchain network regarding the number of wallet creations based on the number of participants. These results are very limited, since only 15 participants were involved, and no other metrics were taken into account.

A blockchain-based system that preserves privacy is presented in Wang and Kogan (2018) for transaction processing systems. The main idea is to devise a trade-off between the advantages of sharing information (transparency) and costs associated with weakened privacy. Privacy protection in blockchain refers to the fact that only authorized parties can access sensitive or proprietary transaction data. The proposed system is based on homomorphic encryption and a recent innovation called zero-knowledge proofs. Only the transaction history is mentioned in the MultiChain tool. Unfortunately, no performance evaluation was provided in order to discuss processing costs, which in our opinion might not be practical for a wide range of data types used in production plants.

4.2. Theoretical foundations of simulation

In order to provide an instructional presentation of our proposal, this section will focus on presenting the basic concepts related to modeling and simulation.

4.2.1. *Definitions*

4.2.1.1. *System*

A system is characterized by the fact that a distinction can be clearly made between what belongs to it and what does not belong thereto. It is assumed that it is controllable and/or observable. Variables generated by the environment act on the system behavior which, in turn, reacts with this environment:

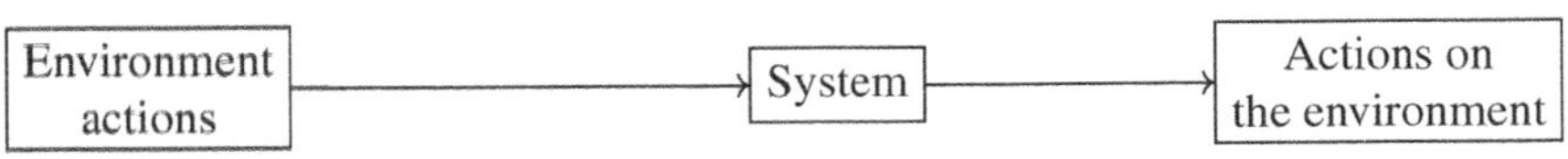

4.2.1.2. *Experimental design*

An experiment is a process by which data about a system are collected by acting on its inputs. The experimentation scheme defines a limited set of circumstances under which the experimentation is conducted:

- observed variables;
- input sequencing;
- initial conditions;
- stop conditions;
- data collection and coding.

4.2.1.3. *Model*

A model M for a system S for an experiment E corresponds to anything to which E can be applied to answer questions about S.

This is therefore a simplified and observable representation of the behavior of the system structure for the purposes of solving an analysis or design problem.

Two types of models can be identified:

– a predictive model to predict a system situation or state;

– a descriptive model where knowledge is capitalized within a model.

4.2.1.4. *Modeling*

Modeling is a process by which knowledge about a given system is organized (Zeigler 1976).

4.2.1.5. *Paradigm*

A paradigm is a set of laws and concepts for the definition of a collection of models:

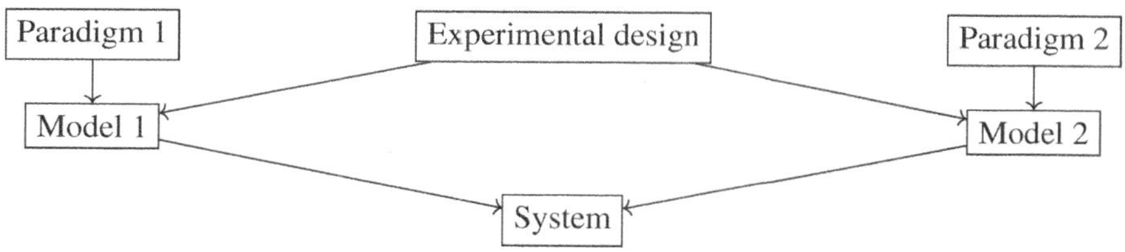

4.2.1.6. *Simulation*

A simulation is a model-based reproduction of the behavior of a real system. It enables one to observe the behavior of a system model over time.

In computer science, it represents the translation of the dynamic behavior of models into a computer program. It is executed in finite time:

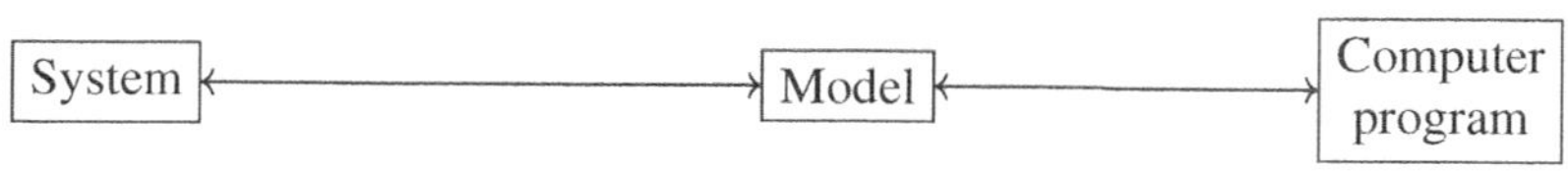

4.2.2. *Model classes*

Model classes differ mainly according to two parameters: time and type of state variables.

4.2.2.1. *Continuous time*

In a continuous-time model, time is specified as continuously changing, it takes the form of a real number.

4.2.2.2. *Discrete time*

In a discrete-time based model, time moves forward by steps from one integer value to another, and it is represented by a whole number.

4.2.2.3. *Discrete state*

In a discrete-state model, variables take their values from a discrete set.

4.2.2.4. *Continuous state*

In a continuous model, the values of the variables are real numbers.

4.2.3. *Specification hierarchy*

In modeling, a hierarchical approach consists of categorizing the known information into different levels. Table 4.1 represents a summary of these different levels as well as the information known to each of them.

Level	Name	Known level elements
0	Observation frame	Variables to be measured and how to observe them on a time base
1	Input/Output relationship	Time-indexed data as well as I/O ports
2	I/O function	Initial state knowledge
3	State transitions	How inputs affect states and how states affect outputs
4	Paired components	How lower level components are coupled

Table 4.1. *Specification hierarchy in modeling and simulation theory*

4.2.4. *Modeling formalisms*

In the modeling process, the choice of formalism represents a strong assumption. This choice has to take system properties into account. Table 4.2 shows formalism examples that can be used based on the characteristics of a model.

For the purpose of the evaluation presented in this book, we will focus on the formalism of discrete event models.

4.2.4.1. *Discrete event system specification (DEVS)*

DEVS is a modeling formalism initiated by Zeigler (1976). It refers to a systemic formalism:

– discrete events;

– state transition sets, states and functions;

– modular and hierarchical approach.

Model characteristics			
State change	Time	Space	Formalism
Continuous	Continuous	Absent	Ordinary differential equations
	Continuous	Continuous, discrete	Ordinary differential equations
	Discrete	Absent	Difference equations
	Discrete	Continuous, discrete	Difference and spatialized equations
Discrete	Continuous	Absent, continuous and discrete	Discrete event models
	Discrete	Absent	Finite state automata
	Discrete	Continuous	Discrete time models
	Discrete	Discrete	Cellular automata

Table 4.2. *Overview of formalism types according to model characteristics*

4.3. Introducing ARTIS*: a DEVS C++ 11 library

The factory 4.0 simulator presented in this chapter was developed based on a framework called ARTIS* (Ramat 2022) developed in C++ 11 and used for the multi-modeling and simulation of complex dynamic systems.

It solves the reliability problem using developments in modeling and simulation theory proposed by Zeigler with the discrete event system specification (DEVS) formalism and its extensions. Everything is defined in the form of template-based classes, more specifically the dynamics of atomic models, the connection graph, parameters, model observations and model serialization. This section will provide a more detailed discussion of some aspects of its technical operation.

4.3.1. *Atomic model*

From a mathematical perspective, an atomic model can be seen as a structure consisting of variables and functions.

DEINITION 4.1.– *Mathematical representation for an atomic model:*

$$DEVS = \langle X, Y, S, S_0, \delta_{int}, \delta_{ext}, \delta_{conf}, \lambda, ta \rangle$$

where:

– *X: all input ports and attached values;*

– *Y: the set of output ports;*

– *S: the set of system states;*

– *δext: external transition function that represents the changes received by the input ports;*

– *δint: internal transition function that represents autonomous changes;*

– *$\delta conf$: conflict function that represents the system responses if external and internal events occur simultaneously;*

– *ta: time during which the model remains in state S;*

– *λ represents external influences.*

In ARTIS*, an atomic model is represented as a class inheriting from artis::pdevs::Dynamics. This class is a template and accepts three parameters:

– the used notion of time;

– the class itself;

– the parameter types accepted by the atomic model.

The atomic model will be initialized by the constructor in terms of parameters. Whether atomic or coupled, any model must be named.

```
struct ExampleParameters {
  double x;
  double y;
};

class Example : public artis::pdevs::Dynamics<artis::common::
                            DoubleTime,
                            Example, ExampleParameters> {
public:
    Example(const std::string &name,
           const artis::pdevs::Context<artis::common::DoubleTime,
                 Machine,
                 ExampleParameters> &context)
      : artis::pdevs::Dynamics<artis::common::DoubleTime,
            Machine,
            ExampleParameters>(name, context) {
        // ...
    }
}
```

Listing 4.1. *Parameter declarations for an atomic model*

4.3.1.1. *Constructor and parameters*

An atomic model includes parameters as data structures. These structures are defined outside of the template code. The parameters are accessible in the model constructor through the context.

4.3.1.2. *Features and methods to be implemented*

4.3.1.2.1. Start

$$DEVS = \langle X, Y, S, \boldsymbol{S_0}, \delta_{int}, \delta_{ext}, \delta_{conf}, \lambda, ta \rangle$$

It is called during the simulation initialization after all the models have been built. The duration of the initial state is defined by the function *ta*.

```
virtual void artis :: pdevs :: Dynamics :: start(
const typename Time :: type& time );
```

Listing 4.2. *Atomic model start method in ARTIS**

4.3.1.2.2. The internal transition function δint

$$DEVS = \langle X, Y, S, S_0, \boldsymbol{\delta_{int}}, \delta_{ext}, \delta_{conf}, \lambda, ta \rangle$$

It is called when *ta* returns and after λ.

```
virtual void artis :: pdevs :: Dynamics :: dint(
const typename Time :: type& time );
```

Listing 4.3. *Atomic model δint method in ARTIS**

4.3.1.2.3. The external transition function δext

$$DEVS = \langle X, Y, S, S_0, \delta_{int}, \boldsymbol{\delta_{ext}}, \delta_{conf}, \lambda, ta \rangle$$

It is called when the model receives at least one external event. The pattern can be changed, but it does not trigger the call to function λ.

```
virtual void dext(const typename Time :: type& t,
const typename Time :: type& e,
const common ::Bag <Time >& bag );
```

Listing 4.4. *Atomic model δext method in ARTIS**

4.3.1.2.4. The conflict function δconf

$$DEVS = \langle X, Y, S, S_0, \delta_{int}, \delta_{ext}, \boldsymbol{\delta_{conf}}, \lambda, ta \rangle$$

It is called when the model receives at least one external event that may have occurred simultaneously with another event. The model can be modified but does not trigger the call to function λ.

```
virtual void dconf(const typename Time :: type& t,
const typename Time :: type& e,
const common ::Bag <Time >& bag );
```

Listing 4.5. *Atomic model δconf method in ARTIS**

4.3.1.2.5. The forward time stepping function ta

$$DEVS = \langle X, Y, S, S_0, \delta_{int}, \delta_{ext}, \delta_{conf}, \lambda, \boldsymbol{ta} \rangle$$

Called after all the transition functions δint, δext or δconf, it returns the amount of time during which the model remains in the state provided by the call to δint, δext or δconf. By default, it returns an infinite time. This is a *const* function so it does not alter the state.

```
virtual typename Time :: type ta(
const typename Time :: type& time) const;
```

Listing 4.6. *Atomic model ta method in ARTIS**

4.3.1.2.6. Lambda λ external influence

$$DEVS = \langle X, Y, S, S_0, \delta_{int}, \delta_{ext}, \delta_{conf}, \boldsymbol{\lambda}, ta \rangle$$

It is called when a *ta* completes and before a δint. The generated external events are returned as a *bag* object. This is also a *const* function, and as such does not alter the model state.

```
virtual common ::Bag <Time > lambda(
const typename Time :: type& time) const;
```

Listing 4.7. *Atomic model λ method in ARTIS**

4.3.1.3. *Attribute declarations*

$$DEVS = \langle X, Y, \boldsymbol{S}, S_0, \delta_{int}, \delta_{ext}, \delta_{conf}, \lambda, ta \rangle$$

Attributes can only be modified inside the functions *start*, δ_{int}, and δ_{ext}. Every type of attribute is possible.

```
// ...
private:
    double _x;
    double _y;
    State _state;
```

Listing 4.8. *Example of attribute declaration in an atomic model*

4.3.1.4. *Definition of input and output ports*

$$DEVS = \langle \boldsymbol{X}, \boldsymbol{Y}, S, S_0, \delta_{int}, \delta_{ext}, \delta_{conf}, \lambda, ta \rangle$$

X and *Y*, respectively, represent the atomic model input and output ports.

Port naming is achieved by enumerations.

```
// ...
public:
  struct inputs {
    enum values { IN_1, IN_2 };
  };

  struct outputs {
    enum values { OUT_1};
  };
```

Listing 4.9. *Port naming in an atomic model*

The declaration of the input and output ports is done in the constructor.

```
// ...in the constructor
input_port({inputs::IN_P, "in_1 "});
output_port({outputs::OUT_P, "out_1 "});
```

Listing 4.10. *Port declaration in an atomic model*

4.3.1.5. *Definition of observables*

A model can be observed at any time and must be able to calculate certain observational variables.

These observation variables are defined by the model.

```
public:
    struct vars {
        enum values {
            NAME,
            STATE
        };
    };
```

Listing 4.11. *Observable naming in an atomic model*

```
// ...in the constructor
observable({vars::NAME, "name"});
observable({vars::STATE, "state"});
```

Listing 4.12. *Observable declaration in an atomic model*

Every atomic model implements an *observe* function to link the name of an observable to the observed value (Listing 4.13).

```
artis::common::event::Value
Example::observe(const artis::factory::Time &,
                 unsigned int index) const {
    switch(index) {
        case vars::NAME:
            return _name;
        case vars::STATE:
            return _state;
    }
    return artis::common::event::Value;
}
```

Listing 4.13. *Implementation of the observed function in an atomic model*

4.3.2. *Coupled model*

A coupled model is a model comprising sub-models that can be atomic or equally coupled. It also includes input and output ports. It is represented as a class inheriting from artis::pdevs::GraphManager.

4.3.3. *Constructor and parameters*

Just as with the atomic model, the parameters of the coupled model are defined in structures outside the model. The difference is in the necessity to define an additional structure encapsulating the parameter structures of the submodels (Listing 4.14).

```
struct SubModelsParameters {
    artis::factory::DispatcherParameters
    ↪ dispatcher_parameters;
    artis::factory::MachineRouterParameters router_parameters;
}

struct GraphManagerParameters {
    std::size_t nb_machines;
}

LineGraphManager(
      artis::common::Coordinator<artis::common::DoubleTime>
          *coordinator,
      const SubModelsParameters &parameters,
      const GraphManagerParameters &graph_parameters)
      : artis::pdevs::GraphManager<artis::common::DoubleTime,
                        LineSubModelsParameters,
                        LineGraphManagerParameters>(
            coordinator, parameters, graph_parameters) {
       // ...
}
```

Listing 4.14. *Example of a coupled model constructor and parameters*

Let us assume that we have a coupled model *LineGraphManager* representing a production line and a structure *GraphManagerParameters* representing the parameters of the coupled model.

The production line includes two submodels, namely the dispatcher and the router, thus requiring the *SubModelsParameters* structure containing the parameters of the *DispatcherParameters* and *RouterParameters* submodels.

4.3.3.1. *Definition of submodels and connections*

The submodels are represented by enumerations to identify them (Listing 4.15).

The first step therefore consists of declaring the submodel in the enumeration.

```
public:
  enum sub_models {
    DISPATCHER,
    ROUTER,
    MACHINE
  };
```

Listing 4.15. *Submodel naming*

The submodel then needs to be added to the coupled model through the *add_child* method in the coupled model constructor (Listing 4.16).

```
this->add_child(DISPATCHER, &_dispatcher);
this->add_child(ROUTER, &_router);
this->add_child(MACHINE, &_machine);
```

Listing 4.16. *Adding a submodel to a coupled model (coordinator)*

The final step in the submodel definition is to link to the coupled model and the other submodels by defining the connections between input and output ports.

Listing 4.17 revisits the example of the production line (coupled model) including two submodels (router and machine). The line input is connected to the router, the router output is connected to the machine input and finally the machine output is connected to the line output.

```
coordinator -> input_port ({IN , "in"});
coordinator -> output_port ({OUT , "out"});
in({ coordinator , IN}) >> in ({&_router, ROUTER::IN });
out ({&_router, ROUTER::OUT }) >> in ({&_machine, MACHINE::IN
  ↪ });
out ({&_machine, MACHINE:: OUT }) >> out ({ coordinator , OUT
  ↪ });
```

Listing 4.17. *Declaration of the connections between the coupled model and submodels*

4.3.3.2. *View definition*

The views are defined to observe the overall model and are connected to outputs (.csv files in ARTIS*).

Technically speaking, a view is a class inheriting from artis::common::observer::View and that includes a set of paths toward the atomic models as attributes. A path consists of the list of identifiers of the coupled models, the atomic model and the observation variable.

Listing 4.18 shows an example of a view for a machine in which the manufacturer expects as parameters the line and machine ID. Paths to atomic models are defined via the *selector* method.

```
class MachineView : public artis::common::observer::View<
        artis::common::DoubleTime> {
public:
  MachineView(int line_id, int machine_id) {
    selector("machine:name", {
        artis::factory::FactoryGraphManager::LINE + line_id,
        artis::factory::LineGraphManager::MACHINE +
          ↪ machine_id,
        artis::factory::Machine::vars::NAME});
    selector("machine:state", {
        artis::factory::FactoryGraphManager::LINE + line_id,
        artis::factory::LineGraphManager::MACHINE +
          ↪ machine_id,
        artis::factory::Machine::vars::STATE});
    }
```

Listing 4.18. *View declaration in ARTIS**

The observable *NAME* is accessed through the coupled model *LINE*, the atomic model *MACHINE* and finally *NAME*, which is an observation variable of *MACHINE*.

4.4. Requirements specification for the proposed evaluation

A comprehensive and relevant presentation of our approach to evaluate the impact of a blockchain-based traceability solution in a factory requires a case study. We shall therefore resort to BPCAT, which was presented in section 3.4 and summarized in Figure 3.22. The blockchain network is made up of three types of blockchain nodes that represent the validator (the company that owns the traceability data), the customers and the service providers involved in the product traceability. Obviously, any additional member of any type can be added as needed. In this blockchain network, a blockchain operates with a copy stored on each node. Product traceability data recorded in the blockchain can be classified into several categories: single-partner data that only relate to the manufacturing plant such as product machining data (Figure 5.22), third-party component integration data that involves suppliers and shipping data involving customers. Some of these data may be confidential.

BPCAT requires certain preconditions on how data should be structured. Each time a new traceability data item is added to a block, a transaction is created. A regular blockchain transaction always includes a transaction hash and a timestamp to ensure that a given action occurred at a specific time in the blockchain. However, since the transaction hash does not guarantee data integrity, BPCAT also adds the data hash into the transaction. With the data hash, timestamp and transaction hash, transaction transparency and data integrity can be guaranteed to the different stakeholders involved.

Confidential data are encrypted using an encryption method chosen by the data owner to ensure confidentiality. Nonetheless, in order to ensure transparency, BPCAT suggests that any encrypted data in a transaction should be included along with further related information that could be used to confirm its authenticity and integrity if necessary. The original data hash before encryption, or the hash of the data derived from the original data in a predefined way, can be used as supporting information. These data are inserted into the transaction, and the timestamp and transaction hash ensure the authenticity and integrity of the whole transaction.

In the event of a dispute, any partner can request the data be decrypted so that they can be compared to the hash stored in the blockchain, and thus establish its authenticity. Encrypted data can only be decrypted by the drive that encrypted it. In addition, digital signatures are used for resolving disputes when a stakeholder's liability is questioned. For each blockchain transaction involving traceability data, each stakeholder (validator, customer, supplier) implicated in the product manufacturing stage must sign the transaction using their personal private key. The signed message contains the transaction hash and a timestamp allowing the signature to be authenticated and guaranteeing non-repudiation.

In order to address issues concerning the amount of traceability data that could involve the use of a cloud storage-based solution, BPCAT does not require storing the files in the blockchain. Each blockchain participant can resort to external storage. The file hash is stored in the blockchain to ensure file integrity and authenticity, as well the file path such that the file can be easily retrieved. This way, any of the players can resort to a cloud storage solution if needed.

This proposal was evaluated based on a prototype using the MultiChain platform, in terms of data and consumed energy volumes (Mullet et al. 2022). However, while the evaluation gives an idea about the impact of blockchain based on selected data, the prototype is not able to evaluate its impact in realistic production scenarios.

4.5. Functional presentation of the simulation model

The simulation model consists of two parts: the overall factory view that can be seen in Figure 4.1, and the production line view is shown in Figure 4.2

4.5.1. *At plant level*

We first find a main factory coupled model used to represent the entire factory. A coupled model is composed of several atomic submodels. For the factory, we find the following elements.

4.5.1.1. *Production order (PO) generator*

This is the input point of the simulator. It is responsible for generating orders and instantiating products with their manufacturing range so that their manufacture can be initiated in the factory. It therefore has only one output port *OUT* for transmitting orders to the next model.

4.5.1.2. *Line*

This is the generic representation of a production line. The factory can contain between 1 and N lines, and there is no predefined limit. It can be a processing, assembly or testing factory. Just like the factory, this is a coupled model that will therefore include several atomic sub-models. These will be described in the section below. The line has an input port *IN P* for receiving products, a port *IN C* for components, an output port *OUT P* for products, a port *OUT D* for component requests and a port *OUT T* for traceability information.

4.5.1.3. *Line router*

Since each atomic model is independent and products have to transit between several production lines during their manufacture, a model capable of transferring a product from one line to another is needed: the model responsible for this task is the line router. When product processing completes on a production line, it is sent back to the router through the line output port *OUT P* to be routed to its next destination.

4.5.1.4. *Global stock*

This is the representation of the factory stock. It includes all the components delivered by the suppliers. These can be routed to the production lines through the request port *IN D*. The component will then leave through the port *OUT C* corresponding to the line. An output port *OUT T* is also included for transmitting traceability information.

4.5.1.5. *Shipping*

This is the final product destination after going through all the destinations comprised in its manufacturing sequence. The final product therefore comes in through port *IN P* and stage-specific traceability information is transmitted through an output port *OUT T*.

4.5.1.6. *Blockchain*

This is the atomic model representing the blockchain and more generally the factory traceability system. In fact, any model capable of generating traceability data includes a port *OUT T* connected to port *IN T* in the blockchain.

4.5.2. *At line level*

As explained earlier, the Line model observed at plant level is a coupled model. It consists of the following submodels.

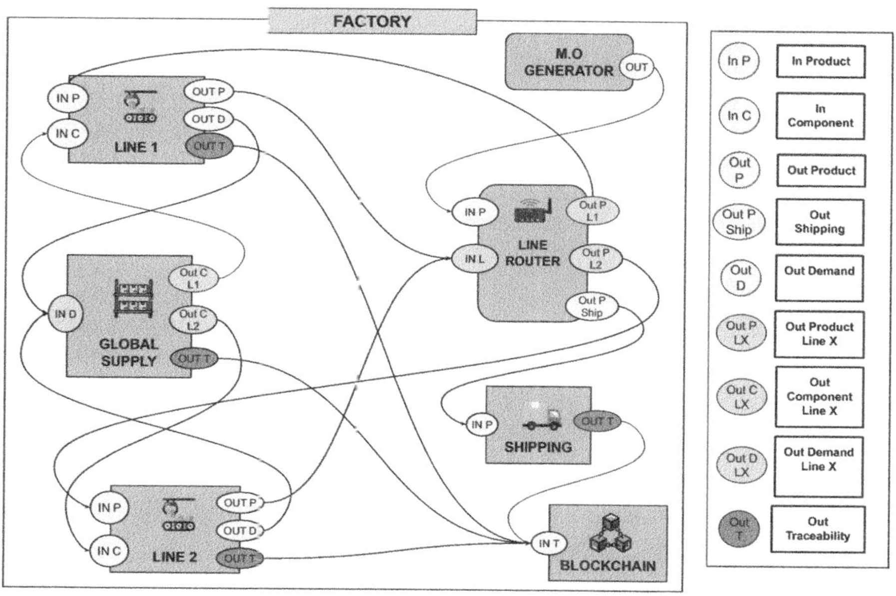

Figure 4.1. *Plant-level simulation model*

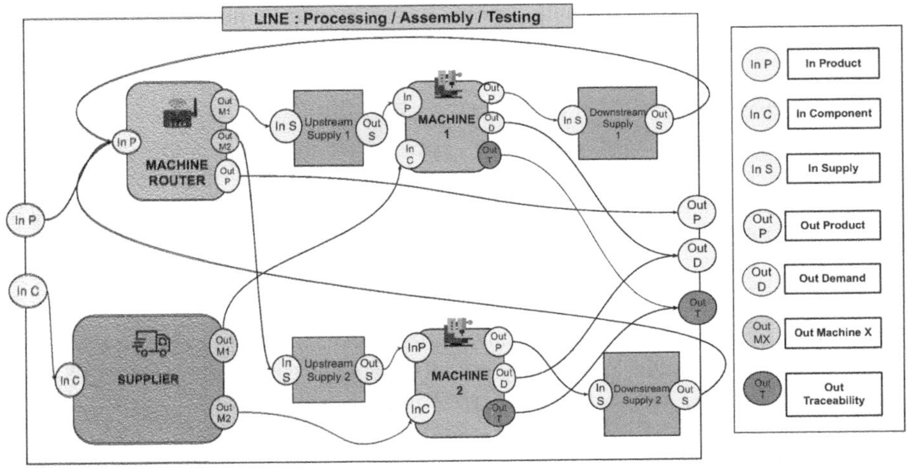

Figure 4.2. *Production line-level simulation model*

4.5.2.1. *Machine*

This is the generic representation of a production machine. A given production line may comprise between 1 and N machines, and there is no preset limit. A machine is expected to receive products through its port *IN P*, but also components from the global stock through port *IN C*. The supply of components is done by means of a request sent by the line to global stock through port *OUT D*. Within a line, the machines are the main element that generate traceability data, since they are in direct contact with the product and perform operations on it. Traceability data are sent through port *OUT T* to the line output.

4.5.2.2. *Upstream supply*

Each machine accesses an upstream supply model representing the products waiting to be processed by the machine. Machines follow a specific operating rate called cycle time that limits the number of products that can be processed during a time period. It therefore only has an input port *IN S* to retrieve products from the waiting list and an output port *OUT S* to send them to the associated machine.

4.5.2.3. *Downstream supply*

As with upstream inventory, each machine is associated with a downstream supply that represents the products processed by the machine and waiting to be sent to their next destination.

4.5.2.4. *Machine router*

Similar to the factory-level line router, the machine router directs products to the next machine in the line because each machine is independent. The router therefore has as many output ports as there are machines in the line (*ports OUT MX*).

4.5.2.5. *Supplier*

Once the line has issued a component request to global supply, the global supply redirects the requested component to the line. Since we know that this component is bound for a machine on the assembly line, we need a model capable of directing the components received by the line to the appropriate machine: this is precisely the role of the supplier. Therefore, the supplier has as many *OUT MX* outputs as line machines, and each of these outputs is connected to the *IN C* input ports of the components for the machines.

4.6. Factory 4.0 simulator technical presentation with ARTIS*

4.6.1. *Architecture*

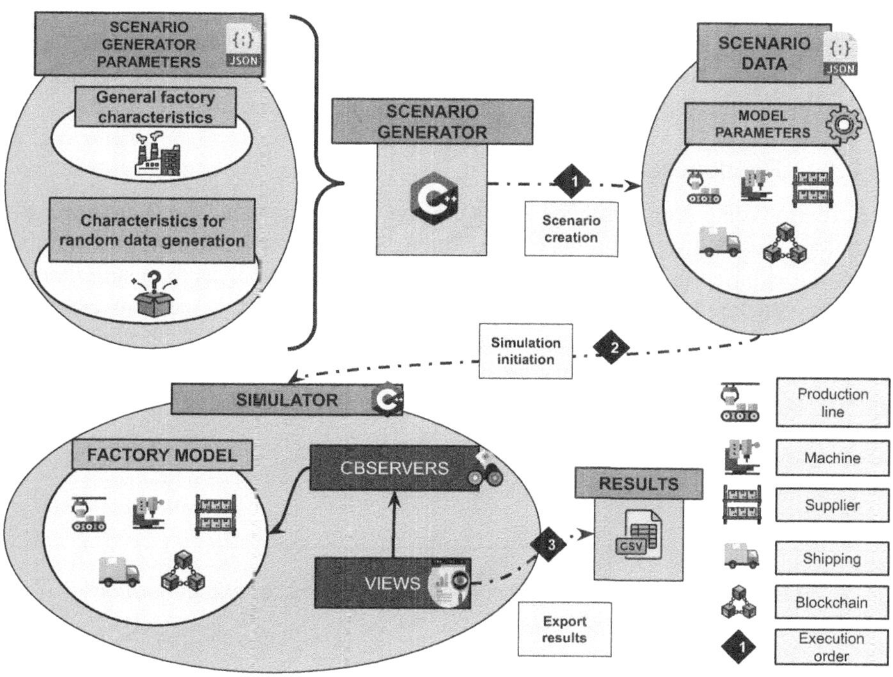

Figure 4.3. *Simulator software architecture for Factory 4.0*

The simulator software architecture described in Figure 4.3 comprises the following elements:

– a scenario generator developed in C++ 11 using a JSON configuration file;

– the JSON file scenario containing the simulation model parameters;

– the simulator developed in C++ 11 comprising the factory model.

In contrast, the simulation results are exported as .csv files.

4.6.2. *Scenario generator*

The structure of the scenario can become rather complicated given that it accounts for the large number of parameters required to model the various elements involved in a plant. Consequently, it made sense to devise a scenario generator capable of creating the complete plant configuration according to various general characteristics. This generator can be parameterized using a *JSON* configuration file.

4.6.2.1. *Generation of pseudo-random values*

In order to be able to generate various scenarios, we will make use of pseudo-random generators to generate numbers. A pseudo-random generator is a deterministic algorithm, that is, a finite and unambiguous sequence of operations. From a given initial number, it will always generate the same sequence of numbers.

4.6.2.1.1. Random seed

Scenario generation must be deterministic; for the same configuration, the generator must always produce the same scenario. Pseudo-random number generation is based on an initial number called *random seed* so as to ensure that the sequence generated will always be the same. This is why the scenario generator configuration includes a parameter *random_seed* to set the value of the random seed and thus ensure the deterministic characteristic.

```
{
    "random_seed": 32456941
}
```

Listing 4.19. *Random seed configuration in the scenario generator*

In C++, we use the Mersenne Twister pseudo-random generator, and more specifically its *MT19937* variant which has the characteristic feature of providing a much longer period in terms of the sequence of numbers that can be generated.

```
uint _random_seed;
std::mt19937 _rng;

_rng = std::mt19937(_random_seed);
```

Listing 4.20. *Example of using the random seed in the C++ scenario generator*

4.6.2.1.2. Probability distributions

```
std::bernoulli_distribution _distrib_data_type;
std::uniform_int_distribution<> _distrib_raw_data_size;
std::uniform_real_distribution<double>
↪  _distrib_file_diff_rate;

```

Listing 4.21. *Example of pseudo-random distributions in C++*

These pseudo-random number generations are based on probability distributions. With this simulator, we mainly use two types of distributions to produce a variety of values in order to get a closer approximation of a real context:

– *Bernoulli distribution* used to produce random Boolean values according to a defined probability;

– integer or real *uniform generation distribution* used to produce random values that are uniformly distributed over a given interval. Uniformity means that the average of the generated values tends toward the interval mean.

4.6.2.2. *Plant structure*

The parameters allowing for describing the structure of a particular plant in simulation scenarios are described in Table 4.3.

4.6.2.3. *Traceability data characteristics*

In the scenario generator, traceability data-related parameters serve to control the parameters of the random generation distributions used (Table 4.4).

4.6.2.4. *Number of nodes in the blockchain*

In order to easily scale up the blockchain architecture, it is also possible to define the number of blockchain nodes for each role.

Name	Description	Example
nb_process_lines	Number of *PROCESS*-type production lines (machining)	40
nb_assembly_lines	Number of *ASSEMBLY*-type production lines (assembly)	20
nb_testing_lines	Number of *TESTING*-type production lines (testing)	20
rng_nb_machines_process_lines	Random generation lower and upper bounds for the number of *PROCESS*-type machines on a line	[1,5]
rng_nb_machines_assembly_lines	Random generation lower and upper bounds for the number of *ASSEMBLY*-type machines on a line	[1,5]
rng_nb_machines_testing_lines	Random generation lower and upper bounds for the number of *TESTING*-type machines on a line	[1,5]
nb_product_refs	Number of different products manufactured by the factory (will influence the number of manufacturing sequences to be generated)	70
nb_stock_components	Number of different references for existing components in stock	100
total_day_production	Number of average products manufactured in one day by the factory	30,000

Table 4.3. *Description of factory-related parameters in the Factory 4.0 scenario generator*

Listing 4.22 shows the configuration for the number of blockchain nodes in the scenario generator.

This is a list of *Role-Quantity* tuples where the role can be *validator*, *customer* or *supplier*.

```
{
    "blockchain_nodes_quantities": [
        ["VALIDATOR", 1],
        ["CUSTOMER", 1],
        ["SUPPLIER", 1]
    ]
}
```

Listing 4.22. *Parameterization of the number of simulated blockchain nodes*

Name	Description	Example
distrib_data_type	Value between 0 and 1 representing the data type distribution (raw or file). For example, a value of 0.6 implies that about 60% of the data will be raw data	0.6
distrib_data_confidentiality	Value between 0 and 1 representing the confidential and non-confidential data distribution. For example, a value of 0.4 implies that about 40% of the data will be confidential	0.4
rng_nb_tracea_data	Lower and upper bounds for the amount of traceability data generated by an atomic model (machine or other)	[2, 10]
rng_raw_data_size	Lower and upper bounds for the size of randomly generated raw data (in bytes)	[128, 1,024]
rng_file_data_size	Lower and upper bounds for the size of randomly generated files (in bytes)	[10,240, 102,400]
rng_file_diff_rate	Lower and upper bounds for the randomly generated difference rate for traceability files	[0.1, 0.3]

Table 4.4. *Description of traceability data-related parameters in the scenario generator of the Factory 4.0 simulator*

4.6.3. *Scenario parameterization*

The scenario forms the cornerstone of the simulator in the sense that it embodies the JSON transcription of the set of parameters of the simulation model.

4.6.3.1. *Simulation context parameters*

Context parameters correspond to simulation time values, namely initialization time, end time and the type of observation step used (Table 4.5). By default, the simulation is carried out with an event-based observation step, that is, any state change in a model induces a measurement of the values of the observables, which provides an accurate overview of how the model changes. In situations where this is not appropriate, for example if we would only like to know the state of the model at the end of the simulation, then a time observation step can be used to observe the model every X seconds.

4.6.3.2. *PO generator parameters*

The PO generator accepts two elements as parameters, namely the product manufacturing sequence list and the list of orders sent to the factory. Together these two elements form what is called the PO, which specifies the quantity to be produced, the expected date, etc.

Name	Description	Example
simu_start	Simulation start time in seconds	0
simu_end	Simulation end time in seconds	86,400 (1 day)
timed_observer_step	No simulation observation. By default, observation is done at the event level (value 0) but a time step can also be defined (observation every X seconds)	$X \geq 0$

Table 4.5. *Description of simulation context parameters*

A manufacturing process is the operating procedure describing the steps necessary to manufacture a product. It is linked to the bill of materials, which is the list and quantity of components to be implemented at the different stages of the manufacturing sequence.

```
{
    "manu_programs": {
        "PRDO": [
          {
            "_line": { "_line_id": 2 },
            "_machines": [
              { "_machine_id": 1 }, { "_machine_id": 2 },
              { "_machine_id": 3 }
            ]
          },
          {
            "_line": { "_line_id": 8 },
            "_machines": [
              { "_machine_id": 1 }, { "_machine_id": 2 }
            ]
          },
          {
            "_line": { "_line_id": 12 },
            "_machines": [
              { "_machine_id": 1 }, { "_machine_id": 2 },
              { "_machine_id": 3 }, { "_machine_id": 4 }
            ]
          }
        ]
    }
}
```

Listing 4.23. *JSON code for the product manufacturing process*

Listing 4.23 shows a snippet of the manufacturing sequence as represented in the JSON scenario submitted to the simulator. Here, a sequence is represented by a list of destinations through which the product must transit during its manufacture. The destination consists of a production line *_line*, as well as the list of machines *_machines* on this line.

4.6.3.3. *Machine parameters*

The machines are associated with an assembly line and are represented by several parameters described in Table 4.6.

Name	Description	Example
machine_id	Machine ID	1
machine_name	Machine name	L1M1
machine_type	Machine type with as possible values: – PROCESS (processing); – ASSEMBLY (assembly); – TESTING (testing)	PROCESS
machining_time	Machine cycle time in seconds	3.6
machine_load_time	Part loading time in seconds	1.2
machine_unload_time	Part unloading time in seconds	1.1
required_compo	Reference of the component(s) that the machine needs to perform its task. This field is only needed for *assembly* machines accordingly with the product-centered traceability vision	C00121
product_tracea_data	Traceability data list that the machine generates each time it processes a part/a product	

Table 4.6. *Description of machine parameters in the Factory 4.0 simulator*

It can be seen that three of these parameters are time related, namely *machining_time* the cycle time, *machine_load_time* the loading time needed to carry out a calibration before starting part processing and *machine_unload_time* the unloading time that enables the machine to forward the part in order to be able to process a new one:

$$T_{tot} = T_{cyc} + T_{load} + T_{unload}$$

where:

– Ttot: total time required for the machine to process the part;

– Tcyc: machine cycle time;

– Tload: time to load the part;

– Tunload: time to unload the part.

4.6.3.4. *Blockchain parameters*

Blockchain parameters fall into two categories: general parameters applied to the entire chain (Table 4.7) as well as node-specific parameters (Table 4.8).

Name	Description	Example
block_overhead_size	Excess space in bytes used by each new block regardless of the amount of data contained in the block	2 048
data_hash_size	Size in bytes of the hashes added to transaction traceability data	32
data_subscribing	List of data-subscribed nodes	
encryption_key_size	Size of the encryption key added to confidential data and inserted into the blockchain	128
file_policy_storage	Storage policy for files	*ONCHAIN*, *OFFCHAIN* or *EXTERNAL*
max_block_size	Maximal block size in bytes	8 388 608 (8MB)
mining_algorithm	Used mining algorithm	PROOF_OF_WORK I NO_POW (Round-Robin)
raw_policy_storage	Storage policy for raw data	*ONCHAIN*, *OFFCHAIN* or *EXTERNAL*
rng_mining_difficulty	Lower and upper bounds for randomly generating a mining difficulty (only for proof of work)	[0.01, 0.04]
signature_size	Size of signatures added to the data and inserted into the blockchain	65
target_adjust_freq	Mining difficulty update frequency in seconds (only for proof of work)	60
target_block_time	Block mining frequency in seconds	30
tx_data_capacity	Maximum number of values stored in a transaction	10
tx_overhead_size	Excess space used with each new transaction regardless of its contents	200

Table 4.7. *Description of general blockchain parameters in the simulator*

4.6.3.4.1. General parameters

The general parameters include several notions such as *file_policy_storage* that enable the storage policy to be modified and *raw_policy_storage*, which makes it possible to modify file or raw data storage mode.

It is also possible to modify the mining algorithm used through the *mining_algorithm* parameter.

Name	Description	Example
name	Identifier to differentiate nodes	v1
can_mine	Boolean indicating whether the node has the right to mine	True | False
cpu_tdp	CPU heat dissipation rate in Watts	95
gpu_tdp	GPU heat dissipation rate in Watts	0 (in this case)
memory_consumption	Memory-related power consumption in Watts	0 (in this case)
nb_cpu	Number of CPUs assigned to the blockchain node	2.0
nb_gpu	Number of GPUs assigned to the blockchain node	0
role	Role assigned to the node	– VALIDATOR – CUSTOMER – SUPPLIER

Table 4.8. *Description of the parameters in a blockchain node in the simulator*

4.6.3.4.2. Node parameters

Node parameters are essentially dedicated to managing their mining behavior with, for instance, *can_mine* that defines whether a node is allowed to mine.

Also directly related to mining, the other parameters are used for evaluating energy consumption, such as *cpu_tdp* or *nb_cpu*.

4.6.3.5. *Traceability data parameters*

Listing 4.24 illustrates how the difference rate between two files is determined.

```
double compareFiles(ifstream &in1, ifstream &in2, const size_t
        chunk_size) {
  //...
  while (remaining) {
    char *buffer1 = new char[chunk_size];
    char *buffer2 = new char[chunk_size];
    size_t size = std::min(chunk_size, remaining);

    in1.read(buffer1, size);
    in2.read(buffer2, size);

    if (0 != memcmp(buffer1, buffer2, size)) {
      nequal_chunks++;
    } else {
      equal_chunks++;
    }
    remaining -= size;
  }
  return ((double)nequal_chunks / (equal_chunks +
     nequal_chunks));
}
```

Listing 4.24. *Function for determining the difference rate between two files in C++*

4.6.3.6. *View parameterization*

At the scenario generator level, views can be parameterized so that they will be used for simulation purposes (Listing 4.25).

```
{
    "view_params": [
    ["BLOCKCHAIN_GENERAL_VIEW", "Blockchain_Global", false, []],
    ["BLOCKCHAIN_NODE_STORAGE_VIEW", "Blockchain_Node_Storage",
        false, []],
    ["SHIPPING_GENERAL_VIEW", "Shipping", false, []]
    ]
}
```

Listing 4.25. *Example of view parameterization in the scenario generator*

There are four parameters: the identifier for the view used in the view generator, the name of the .csv file generated, the view status (active or not) and a list of view-specific parameters.

```
void attach_views() {
    for (auto view_param : _scenario->view_params) {
      ViewType view_type = std::get<0>(view_param);
      std::string view_name = std::get<1>(view_param);
      bool is_enabled = std::get<2>(view_param);
      std::vector<std::string> view_args = std::get<3>
          (view_param);
      if (is_enabled) {
        // ...
        switch (view_type) {
        case POG_GENERAL_VIEW:
          _rc->attachView(view_name, new POGView());
          break;
        case BLOCKCHAIN_GLOBAL_VIEW:
          _rc->attachView(view_name, new BlockchainGlobalView());
          break;
        case BLOCKCHAIN_NODE_STORAGE_VIEW:
          _rc->attachView(view_name,
                new BlockchainNodeStorageView());
          break;
          // ...
        }
      }
    }
}
```

Listing 4.26. *Links between view parameters and the simulator*

These parameters are then used to create the views in the simulation by way a view generator, therefore by calling the main function *attach_views* the views can be attached to the simulator (Listing 4.26).

4.6.3.7. *Description of supported views*

The set of views managed by the simulator are described in Table 4.9.

Name	Observables	Description
POG_GENERAL_VIEW	current_id, po_ref	Production order generator standard view (POG)
GLOBAL_STOCK_GENERAL_VIEW	state, po_id, compo_line, compo_machine, compo_ref	Standard supply view
LINE_ROUTER_GENERAL_VIEW	state, po_id, destination, po_ope_index, po_counter	Line router standard view
SHIPPING_GENERAL_VIEW	po_id, po_counter	Shipping model standard view
MACHINE_ROUTER_GENERAL_VIEW	state, po_id, po_ope_index, destination, po_counter	Machine router standard view
DISPATCHER_GENERAL_VIEW	state, po_id, compo_line, compo_machine, compo_ref	Standard supplier view
MACHINE_GENERAL_VIEW	name, state, po_id, po_counter	Machine model standard view
MACHINE_UPSTOCK_GENERAL_VIEW	state, po_id, is_machine_ready, stock_size	MachineUpStock model standard view
MACHINE_DOWNSTOCK_GENERAL_VIEW	state, po_id, stock_size	MachineDownStock model standard view
PRODUCTION_TRACEA_VIEW	lineX_po_counter, lineX_tracea_size	Line-level traceability statistics view
BLOCKCHAIN_GLOBAL_VIEW	state, length, total_tx, next_mining_time, total_raw_size, total_file_size, total_nochain_size, total_onchain_size, total_offchain_size, total_external_size, total_tx_overhead_total_block data_hash_size, encryption_key_size, signature_size, blockchain_size	Overall view of the blockchain model state
BLOCKCHAIN_NODE_STORAGE_VIEW	nochain_size, validator_size, customer_size, supplier_size, watcher_size	Blockchain node-level storage statistic view

Name	Observables	Description
BLOCKCHAIN_PO_STAT_VIEW	po_id, onchain_raw_confid, onchain_raw_nonconfid, onchain_file_confid, onchain_file_nonconfid (similar for offchain)	Storage statistics view for manufactured product-level traceability
BLOCKCHAIN_TX_DETAIL_VIEW	po_id, sender_name, data_id, data_type, storage_policy, is_confidential, data_size, related_item	View representing in detail transactions submitted to the blockchain
BLOCKCHAIN_BLOCK_STAT_VIEW	index, nb_tx, size, miner, difficulty	View representing blockchain block-related statistics
BLOCKCHAIN_HARDWARE_STAT_VIEW	cpu_usage, avg_cpu_usage, energy_consumption, total_energy_consumption	View representing hardware statistics (CPU, power consumption) for a single blockchain node
BLOCKCHAIN_ALL_NODE_HARDWARE_STAT_VIEW	avg_cpu_usage, total_energy_consumption	View representing overall hardware statistics (average CPU usage, total power consumption) for all blockchain nodes
BLOCKCHAIN_ALL_NODE_HOSTING_STAT_VIEW	total_energy_consumption, total_co2_emission, total_kwh_cost	View representing power consumption and ecology-related statistics (average CPU usage, total power consumption) for all blockchain nodes according to hosting location
BLOCKCHAIN_GLOBAL_HOSTING_STAT_VIEW	total_energy_consumption, total_co2_emission, total_energy_renewable_part, total_kwh_cost	View representing power consumption and ecology-related statistics (average CPU usage, total energy consumption) for all blockchain nodes comparing different hosting locations

Table 4.9. *List of supported views in the simulator*

4.7. Presentation of simulated scenarios

4.7.1. *Used storage space evaluation*

In this section, we will take a look at the simulation of the storage space used by the blockchain and the methods to minimize it.

The prototype developed with MultiChain had shown that this consumption could vary according to several parameters, namely the difference rate between files during off-chain storage, which can decrease the overall blockchain size, but also stream subscription or the lack of it which makes it possible to optimize the storage space individually at the node level.

4.7.1.1. *Traceability data characterization*

Used storage space is an intrinsically data-related notion; this is why characterizing the representation of traceability data within the simulator will be the first stage.

In the simulator, the work does not directly involve the data but the properties related to them. Table 4.10 outlines an exhaustive list of these characteristics.

Property	Description	Example
data_id	Identifier to characterize the data: – LxMy: line x - Machine y; – Cx: component X; – SHIP: shipping	L2M4TD6
diff_rate	Average difference rate between two data elements with the same *data_id*. This is a value between 0 and 1: 0 indicates that the data are entirely identical and 1 indicates that they are completely different	$0 \leq x \leq 1$
is_confidential	Indicates whether the data is confidential	True|False
qty	Quantity used per product	1
type	Data type based on the product-centered approach (file or raw data)	*RAW* or *FILE*
unit_size	Unit data size in bytes	1 024

Table 4.10. *Traceability data characteristics in the simulator*

Equipped with this new information, it is now possible to represent the traceability data in JSON in a scenario submitted to the simulator, as can be seen in Listing 4.27.

```
{
    "data_id": "L2M4TD6",
    "diff_rate": 0.18126772124148832,
    "is_confidential": true,
    "qty": 1,
    "type": "FILE",
    "unit_size": 52571
}
```


Listing 4.27. *Traceability data characterization in a simulated scenario*

4.7.1.2. *Space use according to storage mode*

Based on the previously defined characteristics, we can now establish the formula for calculating the size occupied by a traceability data item under different storage modes: namely on-chain mode where everything is stored without optimization and off-chain mode where optimization is possible.

In on-chain mode, the computation is essentially based on data size and quantity as can be seen:

$$E_{co} = T_{ind} * Q_{te}$$

with:

- E_{co}: total used space;
- T_{ind}: individual data element size;
- Q_{te}: used amount per product.

In off-chain mode, duplicate data are not stored; therefore, only differences are kept.

The difference rate thus has to be integrated into the formula as follows:

$$E_{co} = T_{ind} * Q_{te} * T_{diff}$$

where:

- E_{co}: total used space;
- T_{ind}: individual data element size;
- Q_{te}: used amount per product;
- Q_{diff}: difference rate associated with these data.

Listing 4.28 introduces the function *get_consumed_space,* which in practice is the C++ implementation of the two formulas used above during the simulation.

```
std::size_t get_consumed_space(TraceaData *tdata) const {
    PolicyStorage::values ps = get_policy_storage(tdata);
    switch (ps) {
    case PolicyStorage::ONCHAIN:
      return tdata->unit_size * tdata->qty;
    case PolicyStorage::OFFCHAIN:
        // If the datum has already been inserted at least
        // once, we save space
      if (_present_data_ids.count(tdata->data_id) > 0) {
        return tdata->unit_size * tdata->qty * tdata->diff_rate;
      } else {
        return tdata->unit_size * tdata->qty;
      }
      break;
    case PolicyStorage::EXTERNAL:
      return tdata->unit_size * tdata->qty;
    default:
      return tdata->unit_size * tdata->qty;
    }
}
```

Listing 4.28. *Function for computing the amount of space used by a traceability data element according to the storage mode*

4.7.1.3. *Extra cost related to space consumption (hash, signature, etc.)*

Data size is not the only factor to be taken into consideration when evaluating the amount of storage space used.

In fact, when using blockchain for product-centric traceability, further information is actually integrated such as the data hash to guarantee integrity, the encryption key for confidentiality and finally the signature for non-repudiation.

These three elements are taken into account and also added to the scenario configuration (Listing 4.29).

```
{
    "data_hash_size": 32,
    "encryption_key_size": 128,
    "signature_size": 65,
    "block_overhead_size": 2048,
    "tx_overhead_size": 200,
    "raw_policy_storage": "ONCHAIN",
    "file_policy_storage": "OFFCHAIN"
}
```

Listing 4.29. *Configuration of additional elements consuming storage space in the JSON scenario*

It is estimated that the hash size is 32 bits, because the algorithm used is *SHA-256* that encodes the hash in 256 bits, or 32 bytes. The encryption key is 128-bytes long, because the algorithm used is RSA 1024 therefore the key is encoded in 1024 bits or 128 bytes. The signature size is specific to MultiChain (64 bytes + 1 byte).

Finally, it is important to note that a blockchain solution may have to generate additional data to ensure its proper operation. This aspect is taken into account with the values *block_overhead_size*, which represent the additional cost of space used with each new block, and *tx_overhead_size,* which is the additional transaction cost.

4.7.1.4. *Characterization of the notion of subscription*

The second way to reduce the storage space used is based on the notion of subscription which, in MultiChain, applies to streams and makes it possible for a node to define the data it wants to store. Streams are not represented in the simulator in the sense that they can be used to categorize the data; yet, we already have similar information in the data characterization (the *data_id* field).

This is why the subscription will be represented in the scenario by associating each data type with a list of nodes that will be subscribed to them, as can be seen in Listing 4.30. The decisions about subscriptions in the scenarios were made along the lines of the vision of product-centric traceability; in other words, the validator is subscribed to all data types, because it is the one responsible for submitting them to the blockchain. On the other hand, suppliers subscribe to the data type *COTD0,* which is a traceability originating from a component that they have manufactured themselves. The customer subscribes to the *SHIPTD1* type data, which are used by the shipping department.

```
{
    "data_subscribing": {
        "COTDO": ["VALIDATOR", "SUPPLIER"],
        "L6M1TD3": ["VALIDATOR"],
        "L1M2TD4": ["VALIDATOR", "CUSTOMER", "SUPPLIER"],
        "SHIPTD1": ["VALIDATOR", "CUSTOMER"]
    }
}
```

Listing 4.30. *Characterization of the notion of subscription in a simulation scenario*

In the simulation itself, implementing the subscription consists, for a traceability data item, in scanning the list of nodes subscribed to the category of the data item in question and then in adding the data only to the nodes that are subscribed to it, which will therefore increase the storage space used by the nodes.

```
void update_node_stat(TraceaData *tdata) {
  // For each node "subscribed" to this data category
  for (auto cnode : _data_subscribing[tdata->data_id]) {
      // The data are added and the storage
      // space used by this node is increased
      add_data(&_node_tracea_stat[cnode], tdata);
      add_hash(&_node_tracea_stat[cnode]);
      if (tdata->is_confidential) {
          add_encryption_key(&_node_tracea_stat[cnode]);
      }
      add_signature(&_node_tracea_stat[cnode]);
  }
}
```

Listing 4.31. *Implementation of the notion of subscription in C++ in the simulator*

4.7.1.5. *Simulation results*

With this implementation, we performed several simulations in order to demonstrate how the difference rate between traceability files influences the storage space used, as seen in Figure 4.4. This graph therefore presents five different scenarios:

– 100%: all files differ from each other. This is the typical scenario where off-chain mode cannot introduce any improvement.

– 70–90%: average difference rate between 70% and 90%.

– 50–70%: average difference rate between 50% and 70%.

– 30–50%: average difference rate between 30% and 50%.

– 10–30%: average difference rate between 10% and 30%.

For each of these scenarios, four statistics are indicated: the total amount of data inserted that does not change between each scenario, and the storage space used on the validator, client and supplier nodes.

First, it can be seen that when the difference rate between files decreases, the space consumed by the blockchain also decreases, decreasing from 74.36 Gb for the 100% scenario to 15.60 Gb, which shows that it is possible to significantly reduce data storage space.

In addition, if we individually consider each scenario, the benefits of subscribing to data can be clearly seen, since the space occupied on the validator, client and supplier nodes is not the same. The validator node stores all the data, which means that it utilizes a much larger amount of space, while the supplier and customer utilize less space. This is all the more striking since here the supplier and customer nodes, respectively, represent all suppliers and all customers. The actual quantity stored individually by each supplier and each customer would then represent only a fraction corresponding to the components dispatched by that supplier or the products delivered to that customer.

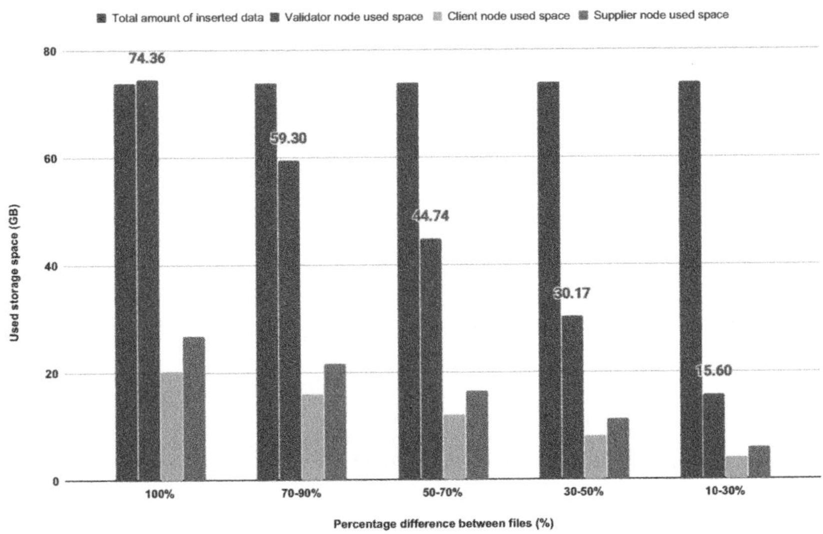

Figure 4.4. *Simulation showing how the storage space used changes according to the difference rate between files in off-chain mode. Inserted data total quality: 73.7Gb*

4.7.2. *Energy consumption assessment*

In the context of the simulation, the energy consumption that we want to assess is that of the blockchain and more specifically the differences according to the mining algorithm used. The evaluation of this consumption consists of evaluating node consumption and consequently that of their computer components.

4.7.2.1. *Determination of the computation formula*

First, let us consider the following computation formula proposed by Microsoft for measuring the electricity consumption of its cloud infrastructure (Bergman 2020):

$$P[kWh] = \frac{c * Pc + Pr + g * Pg}{1000}$$

with:

– c: the number of CPUs;

– Pc: CPU consumption (W);

– Pr: memory consumption (W);

– g: the number of GPUs;

– Pg: GPU consumption (W).

Since there are no node-level graphic computations performed, the GPU consumption is considered to be 0. Also, because memory consumption is very low, it is considered negligible. Node energy consumption will therefore be reduced to the CPU consumption Pc, which can be represented with the following formula:

$$Pc = tdpc * rc * th$$

with:

– $tdpc$: CPU TDP (*Thermal Design Power*) (W);

– rc: CPU utilization rate (in %);

– th: elapsed time (h).

DEINITION 4.2.– *Formula for computing power consumption in a node P in the simulation context is given as:*

$$P[kWh] = \frac{c * (tdpc * rc * th)}{1000}$$

with:

– c: number of CPUs;

– *tdpc: CPU TDP (W);*

– *rc: CPU utilization rate (%);*

– *th: elapsed time (in hours).*

The various parameters of this formula include the number of CPUs *c* and its TDP *tdpc,* which are part of the technical characteristics of the machine on which the blockchain runs. *tdpc* will therefore have a value of 95 W. The number of CPUs allocated to each node will be the same and will be defined at the Docker container configuration level with a value of 2.0, as indicated in Listing 4.32.

```
validator1-node-chain:
    deploy:
      resources:
        limits:
          cpus: '2.0'
          memory: 2gb
```

Listing 4.32. *Configuration of node CPU resources with Docker*

Variables *c* and *tdpc* are now known fixed values; as such, they can be directly integrated into the JSON scenario configuration as can be seen in Listing 4.33.

```
{
"VALIDATOR": {
        "can_mine": true,
        "cpu_tdp": 95,
        "gpu_tdp": 0,
        "memory_consumption": 0.001,
        "nb_cpu": 2,
        "nb_gpu": 0
    }
}
```

Listing 4.33. *Validator node configuration in the scenario in JSON for computing power consumption*

In contrast, the utilization rate *rc* and the elapsed time *th* are dynamic parameters that vary depending on the workload allocated to the node. The next step will therefore consist of determining the formulas that will enable these two values to be predicted.

4.7.2.2. *CPU utilization rate prediction*

This step involves finding a correlation between the CPU utilization rate and the indicators that can be measured in a real blockchain system, such as data and transaction size, for instance. Using the prototype developed as part of the implementation with MultiChain, it was possible to perform node-level measurements according to various parameters:

– the backup mode: *on-chain* or *off-chain*;

– the mining algorithm employed: proof of work or round-robin;

– whether mining is permitted or not.

In order to determine whether a tendency could be observed between the CPU rate and these parameters, a regression was applied to derive an equation. The fit of the regression line to the data points was evaluated by the R^2 value.

For example, Figure 4.5 features the relationship between CPU rate and data size in round-robin where all nodes have the right to mine.

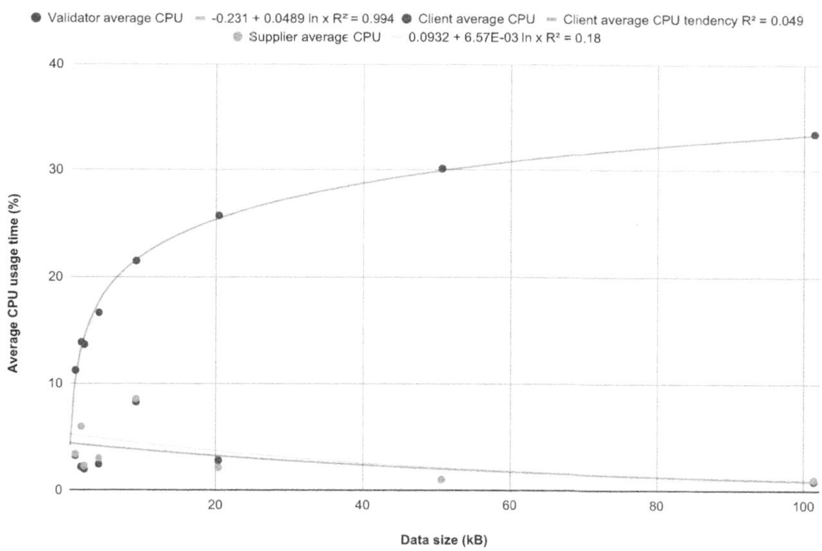

Figure 4.5. *Relationship between CPU rate and data size for round-robin (all nodes are mining)*

A trend can be clearly seen in particular for the validator CPU rate where the regression curve is logarithmic with an R^2 equal to 0.994, which indeed confirms the correlation.

In contrast, for the client and supplier nodes, no clear trend is observed and the CPU rate remains nearly constant, which seems to make sense since their actions merely consist of locally retrieving the transaction data transmitted by the validator and signing the blocks during the mining. Yet, these actions are not very greedy in terms of CPU as shown by their curves. Most of the work, namely tasks that involve recording and processing transactions, is performed by the validator that exhibits the highest CPU activity.

Figure 4.6 shows a similar study, but this time with proof of work.

At first, the measurements were achieved just as for the previous graph, that is, with respect to data size. However, it became visible that there was no strong link between data size and CPU rate with proof of work, as the R^2 value was less than 0.1. Another parameter had to be found, and mining difficulty turned out to be the obvious candidate.

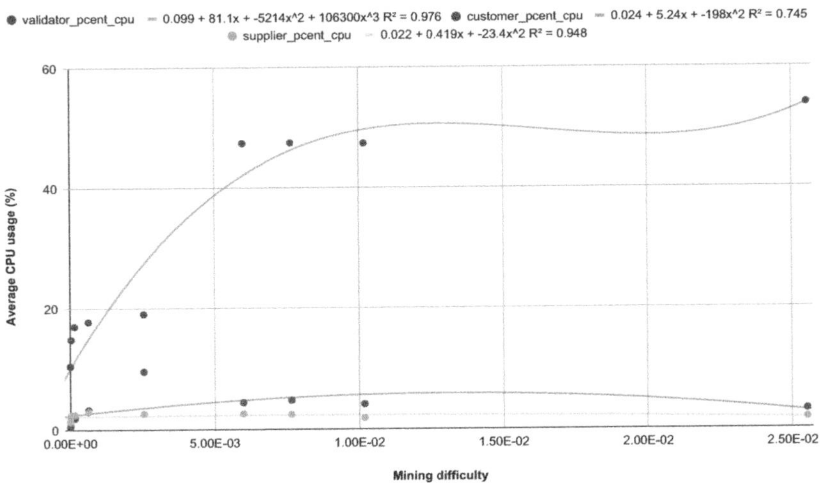

Figure 4.6. *Relationship between CPU rate and data size for proof of work (all nodes are mining)*

Mining difficulty is an arbitrary measure that the blockchain adjusts in order to guarantee a stable block mining time (one block generated every N seconds). This is a value that directly affects the difficulty of the computation and therefore the consumption in terms of CPU resources, hence its choice for establishing this correlation. Given that proof of work is a resource-intensive algorithm, it is important to make a distinction between measurements when the node is the block creator and those when it is not. Figure 4.6, for example, shows the measurements when the validator created the blocks, which explains why its CPU consumption is higher than that of the client and supplier nodes. A strong correlation can be observed with the mining difficulty, because the R^2 values are greater than 0.95 for two of the nodes.

This thus validates the prior assumption where the mining difficulty was chosen to be the parameter to be related to the CPU rate utilized.

The graphs related to the computation of the CPU rate for the client and supplier nodes are shown in Figures 4.7 to 4.11.

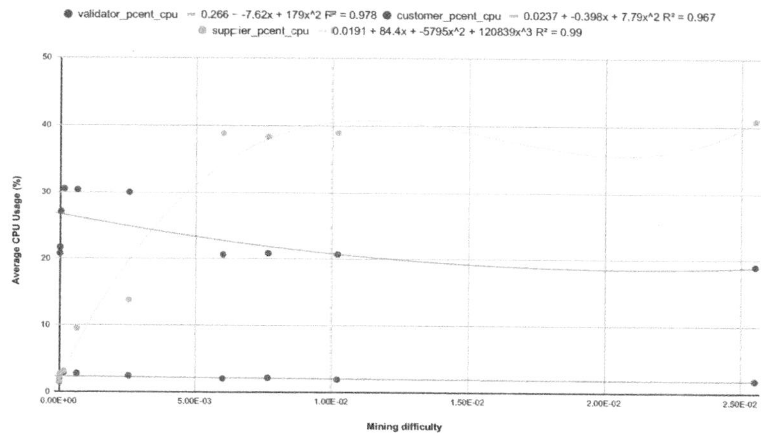

Figure 4.7. *Relationship between CPU rate and mining difficulty for proof of work (all nodes are mining, the supplier is the creator)*

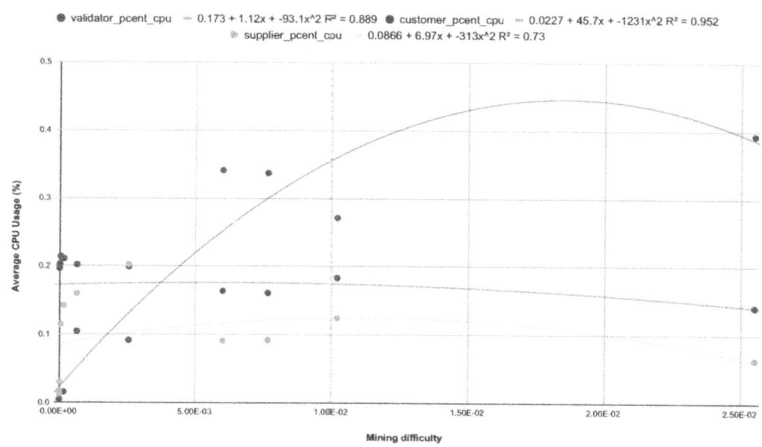

Figure 4.8. *Relationship between CPU rate and mining difficulty for proof of work (all nodes are mining, the client is the creator)*

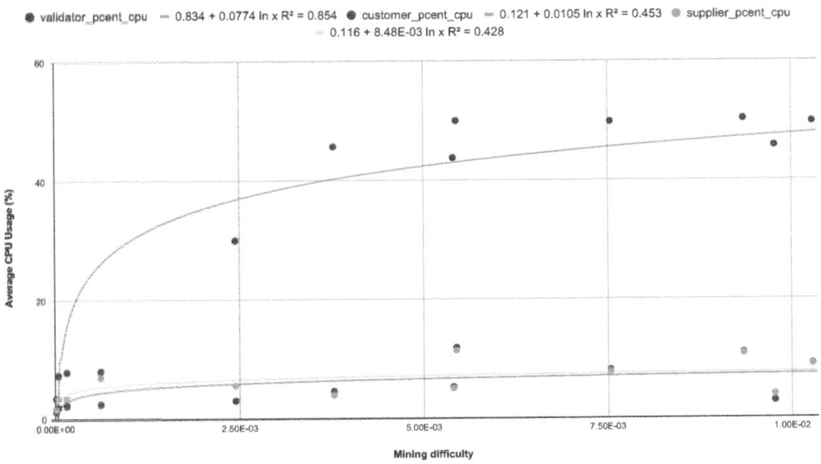

Figure 4.9. *Relationship between CPU rate and mining difficulty for proof of work in off-chain mode (all nodes are mining, the validator is the creator)*

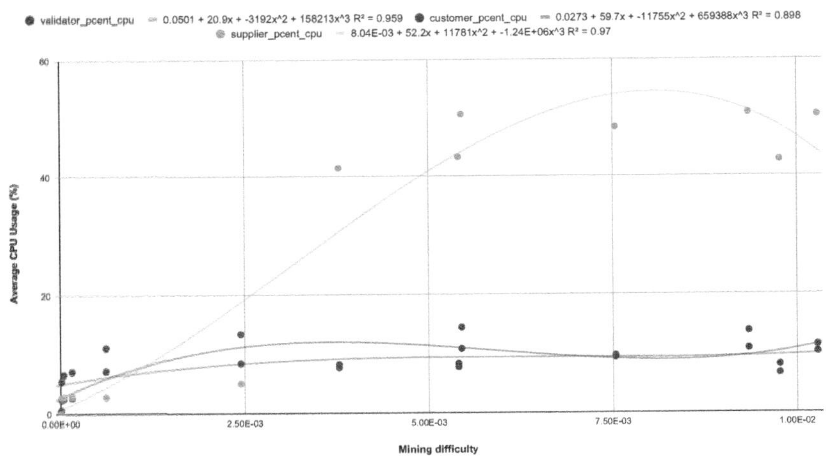

Figure 4.10. *Relationship between CPU rate and mining difficulty for proof of work in off-chain mode (all nodes are mining, the supplier is the creator)*

Once the equations are determined, they must be included in the simulator scenario. Listing 4.34 shows how the equations from the previous graphs are integrated into the JSON scenario used by the simulator. There are different types of equations, namely logarithmic equations or N-degree polynomials.

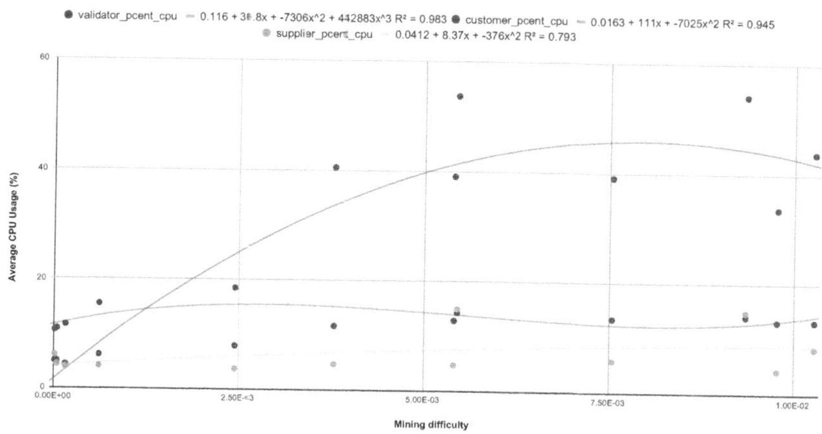

Figure 4.11. *Relationship between CPU rate and mining difficulty for proof of work in off-chain mode (all nodes are mining, the client is the creator)*

```
{
      "cpu_usage_nonpow_onchain_mine": {
        "type": "LOGARITHMIC",
        "coeffs": [0.0489, -0.231]
      },
      "cpu_usage_nonpow_onchain_nonmine": {
        "type": "CONSTANT",
        "coeffs": [0.02]
      },
      "cpu_usage_pow_onchain_mine_multi_validator_turn": {
        "type": "POLYNOMIAL",
        "coeffs": [0.099, 81.1, -5214, 106300]
      }
}
```

Listing 4.34. *Example of integrating equations for the computation of CPU rate in the JSON scenario*

4.7.2.3. *Prediction of transaction processing time*

The second unknown in the computational formula for energy consumption is the transaction processing time *tc*. Similarly as for the CPU rate, processing time measurements were made based on different scenarios. Measuring the transaction processing time consists of calculating the running time T of the function responsible for submitting the data; the calculation can be represented by the following equation:

$$T_{ec} = T_{end} - T_{sta}$$

where:

– T_{ec}: the elapsed time;

– T_{end}: final time when the function terminates execution;

– T_{sta}: initial time when the function starts execution.

The implementation of this formula in Python in the prototype can be seen in Listing 4.35. The *process_time_ns* function does not include elapsed time during idle CPU phases.

```
def time_func(func, **kwargs):
    start = time.process_time_ns()
    result = func(**kwargs)
    elapsed = time.process_time_ns() - start
    return elapsed, result
```

Listing 4.35. *Function for the calculation of the running time of a function*

Since transaction processing is linked to its registration, it would seem logical that time should be linked to data size. Figure 4.12 shows the correlation between processing time and data size in *on-chain* transactions. A clear tendency can be observed that enables us to model time evolution according to a linear equation with a R^2 greater than 0.99.

After determining the equation, it can now be integrated into the JSON configuration of the scenario of Listing 4.36.

```
{
    "processing_time_onchain": {
      "type": "LINEAR",
      "coeffs": [9.22e-7, 2.82e-3]
    }
}
```

Listing 4.36. *Integration of the equation for the calculation of processing time in the JSON scenario*

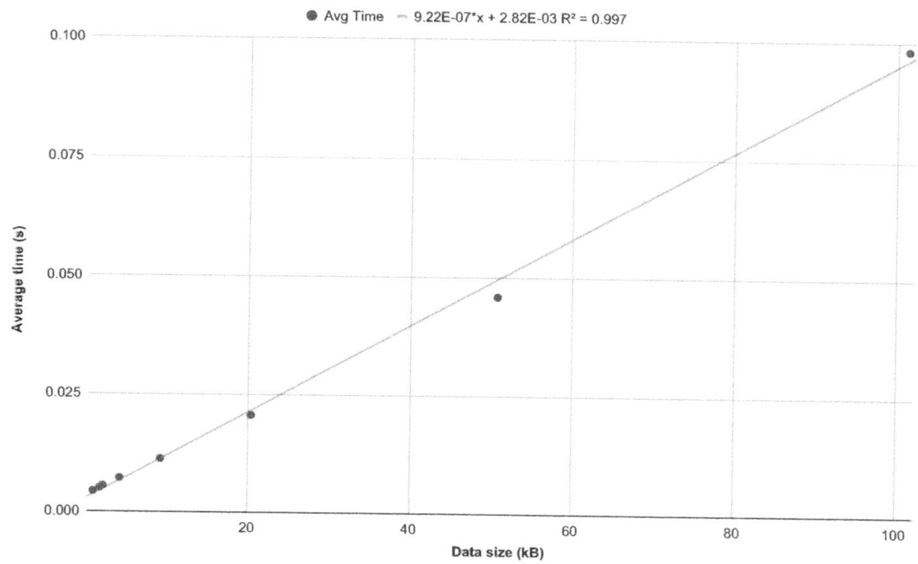

Figure 4.12. *Relationship between transaction processing time and data size (on-chain transactions)*

4.7.2.4. *Integration of the formula for the calculation of power consumption in the simulation*

In the simulation, the calculation of power consumption is implemented as follows (Listing 4.37). The CPU usage rate *cpu_usage* and the transaction processing time *tx_duration* are retrieved and then the equation of definition 4.2 is applied.

```
for (auto node : _blockchain_nodes) {
        double cpu_usage = get_cpu_usage(tdata, &node);
        double tx_duration = get_processing_time(tdata);
        // Formula for computing electric consumption
        double energy_consumption = (node.profile.nb_cpu *
                                node.profile.cpu_tdp *
                                (tx_duration * cpu_usage)) / 1000.0;
}
```

Listing 4.37. *Integration of the formula for the calculation of power consumption in a node in C++*

4.7.2.5. *Comparison of results under different scenarios*

Figure 4.13 summarizes several simulations for the purposes of evaluating blockchain electricity consumption over a 1-day period based on the following scenarios:

- nPOW–VCP (round-robin – all nodes are mining);
- POW–VCP (proof of work – all nodes are mining);
- POW–VP (proof of work – only the validator and the vendor are mining);
- POW–V (proof of work – only the validator is mining).

For each of these scenarios, we have consumption per node (validator, client, supplier) as well as overall consumption that corresponds to all the nodes combined.

First, a very large difference in consumption can be observed between the scenarios using proof of work (1.44 kWh for POW–VCP) and the scenario using the round-robin algorithm (0.27 kWh for nPOW–VCP). The proof-of-work overconsumption in terms of electrical energy is thus confirmed by the simulation.

Across all the scenarios involving proof of work combined, a distribution phenomenon or pattern of power consumption can be perceived as the number of nodes allowed to mine increases: for example, in the POW–V scenario, the validator is responsible for nearly all of the power consumption for a total of 1.71 kWh, whereas in POW–VP, a balanced consumption starts to take shape between the validator and the supplier with 0.81 kWh and 0.62 kWh, which makes sense because they are the only ones authorized to mine. Finally, in the POW–VCP scenario in which every node is mining, the distribution is virtually homogeneous between the three nodes with 0.57 kWh for the validator and 0.44 kWh for the other two.

This is an interesting observation because it can be used as a basis for a discussion on cost distribution between stakeholders in terms of blockchain mining-related electricity consumption. More specifically, insofar as the transparency and integrity that blockchain offers could be seen as services to the factory, its suppliers and its customers, it is suggested here that possibly all players involved should bear the energy cost of its operation, for example, in the form of a subscription.

4.7.3. *Carbon (Co_2) emissions evaluation*

The ecological aspect is one of the challenges of blockchain technology, as mentioned in the previous chapters. For the evaluation of this aspect, measurements of carbon emissions are a solid basis. As a reminder, at this stage, we are already able to estimate the electrical energy consumption of the blockchain using the simulator.

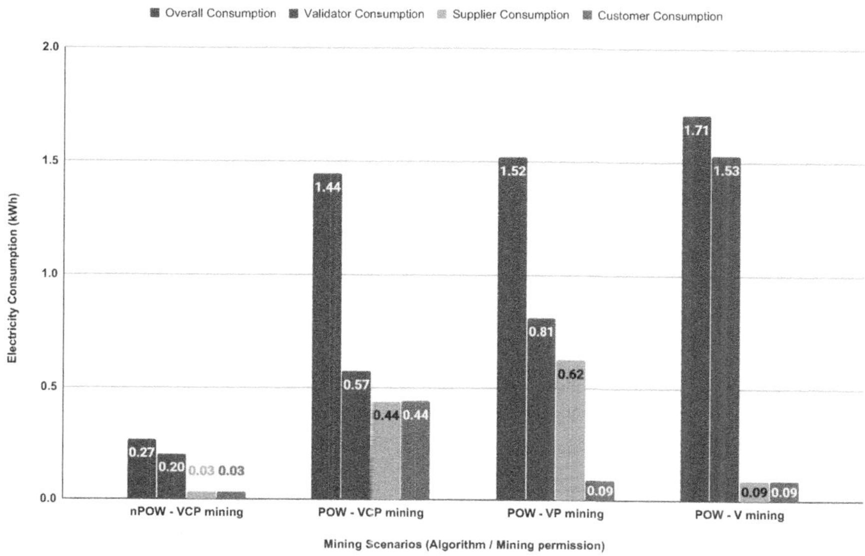

Figure 4.13. *Evaluation of power consumption in blockchain nodes according to different mining scenarios. Nodes: 1x validator, 1x client, 1x supplier, simulation time: 1 day*

4.7.3.1. *Determination of the computation formula*

The electricitymap.org *open source* API allows us to obtain a large number of parameters related to electricity consumption and more specifically carbon intensity by country. Carbon intensity corresponds to the ratio of CO_2 emissions to electricity consumption and is expressed in gCO_2eq/kWh. The data of this API typically originates from grid operators, balancing entities, or market operators linked to the energy market and are used by cloud operators to reduce their carbon footprint.

From the carbon intensity, the quantity of CO_2 emitted E can therefore be determined with respect to consumption P by the following formula.

DEINITION 4.3.– *Formula used for the calculation of the quantity E of CO_2 emitted by a node within the context of the simulation:*

$$E = P * Ic$$

with:

– *E: CO_2 emitted quantity (g);*

– *P: node power consumption (kWh);*

– *Ic: carbon intensity of the country where the node is hosted (gCO_2 eq/kWh).*

Carbon intensity Ic is a known data element with a fixed value, it thus can be integrated into the JSON scenario as shown in Listing 4.38.

```
{
    "hosting_areas":[{
        "area_name":"FR",
        "carbon_intensity":73,
        "renewable_part":0.3,
        "kwh_cost":0.177
      }]
}
```

Listing 4.38. *Parameter configuration to calculate CO_2 emissions in the JSON scenario*

At the simulator level, the computation formula is integrated in C++ as shown in Listing 4.39.

```
for (auto node : _blockchain_nodes) {
      double cpu_usage = get_cpu_usage(tdata, &node);
      double tx_duration = get_processing_time(tdata);
      double energy_consumption = (node.profile.nb_cpu * node.
                           profile.cpu_tdp *
                           (tx_duration * cpu_usage)) /
                           ↪  1000.0;
    for (auto &area : _hosting_areas) {
        // Formula for calculating amount of CO2 emitted
        double co2_emission = energy_consumption *
                       area.carbon_intensity;
    }
}
```

Listing 4.39. *Computation formula for node-level CO_2 emissions in C++*

4.7.3.2. *Comparison of results under different scenarios*

Simulation-related results of CO_2 emissions can be found in the two graphs in Figures 4.14 and 4.15. In both figures, the duration of the simulation is 1 day and the scenarios presented are as follows:

– nPOW–VCP (round-robin – 1 validator, 1 customer and 1 supplier);
– nPOW–VC20P5 (round-robin – 1 validator, 20 customers and 5 suppliers);
– POW–VCP (proof of work – 1 validator, 1 customer and 1 supplier);
– POW–VC20P5 (proof of work – 1 validator, 20 customers and 5 suppliers);
– nPOW–VC100P20 (round-robin – 1 validator, 100 customers and 20 suppliers);
– POW–VC100P20 (proof of work – 1 validator, 100 customers and 20 suppliers).

All nodes are mining in all scenarios.

Figure 4.14 shows the details of CO_2 emissions by node role, the possible roles being validator, client, and supplier. Several nodes can have the same role: for example, the POW scenario – VC20P5 includes 20 customers and five suppliers. This means that customer role emissions correspond to that of the 20 cumulative customer nodes and the consumption of the supplier role to that of the five cumulative supplier nodes.

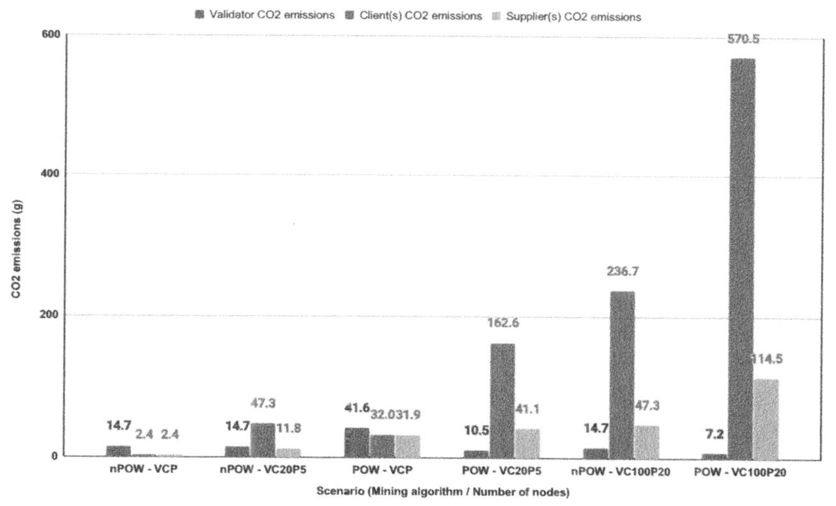

Figure 4.14. *Blockchain CO_2 emissions per node according to the number of nodes and the mining algorithm employed. Hosting country: France (carbon intensity: 73 gCO_2eq/kWh; source: electricitymap.org)*

Initially, it is observed that the scenario emitting the least CO_2 is nPOW–VCP, which is an expected result because it uses the least resource-greedy algorithm – round-robin in this case – while involving the smallest number of nodes (one node for each role).

For all nPOW scenarios, validator emissions remain stable at 14.7 g because the number of validator nodes does not change. On the other hand, customer and supplier role emissions continue to increase because their number is constantly increasing, with the highest value being that of customers in the nPOW–VC100P20 scenario with 236.7 g. It is important to also note a proportional emission distribution between customers and suppliers: for example, in nPOW–VC100P20, the emissions of 100 customers are 236.7 g and those of 20 suppliers are 47.3 g, which corresponds to five times less than the total accumulated by the customers.

With the POW scenarios, a decreasing phenomenon in CO_2 emissions can be observed at the validator level: 41.6 g in POW–VCP and 7.2 g in POW–VC100P20. This is because CO_2 emissions are mainly created by mining-related activity and therefore caused by proof of work. Nonetheless, for increasing numbers of client and supplier nodes, the probability of block mining will decrease for the validator: for example, in the POW–VCP scenario, there is a one in three chance of mining for the validator, while in POW–VC100P20, the validator has a 1 in 121 chance. At the customer and supplier level, emissions increase as the number of nodes increases, and the highest value is emitted by the customers in the POW scenario–VC100P20 with 570.5 g.

Finally, by comparing the scenarios using proof of work (POW) and the nPOW round-robin-based scenarios, it can be seen that for the same number of nodes, proof of work always emits more CO_2 than round-robin.

Figure 4.15 depicts both the total amount of CO_2 emitted by the blockchain, that is, emitted by all nodes, as well as the total electricity consumption, in order to demonstrate that the correlation between the two parameters is accurately accounted for in the simulation. The scenarios are presented in ascending order in terms of power consumption and CO_2 emissions.

Of the six scenarios, we can see that the two most economical are those using round-robin, namely nPOW–VCP with 19.5 g and nPOW–VC20P5 with 73.9. Nonetheless, one of the scenarios using round-robin manages to reach second place: namely nPOW–VC100P20 with 100 customers, 20 suppliers and a total consumption of 298.8 g. The number of nodes should therefore not be overlooked concerning its influence in increasing consumption.

Between nPOW–VCP and POW–VCP, the total consumption increases from 19.5 to 73.9 g when swapping mining algorithms, while between nPOW–VCP and nPOW–VC20P5 the consumption increases from 19.5 to 105.4 g by changing the number of nodes (from 3 nodes to 26). It can be deduced that the influence of the mining algorithm is greater than that of the number of nodes.

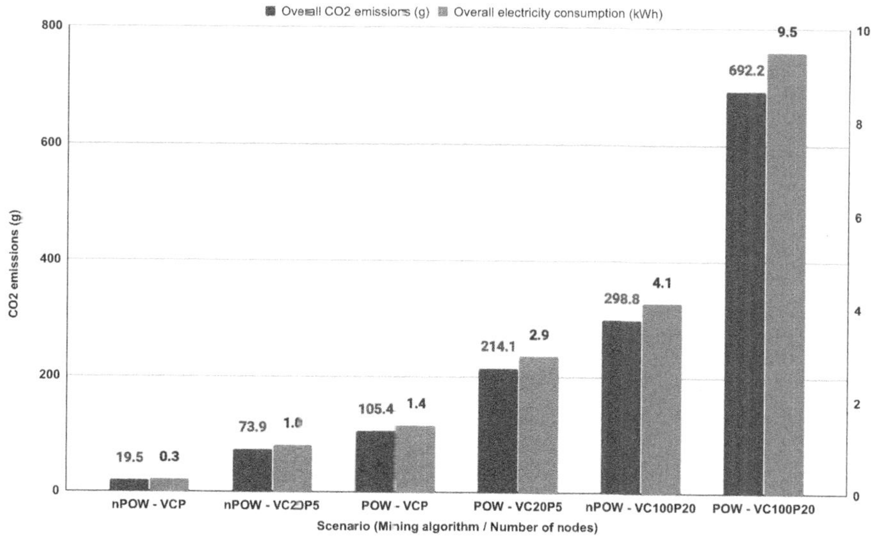

Figure 4.15. *Global blockchain CO_2 emissions based on the number of nodes and the mining algorithm employed. Hosting country: France (carbon intensity: 73 gCO_2eq/kWh; source: electricitymap.org*

4.8. Summary of the strengths and weaknesses of the contribution

In this chapter, we have proposed an evaluation approach based on the discrete-event system specification for Industry 4.0. Our main proposal is a framework for collaborative modeling of a manufacturing plant, in which each component can be as finely modeled and configured as needed by an expert according to the purposes of the evaluation. The framework also offers a general overview of the plant where managers can configure (production planning, storage management, etc.) and observe how models behave (storage volume, energy consumption, etc.).

Applying this approach to our recent proposal for a blockchain-based traceability system in an Industry 4.0 manufacturing plant (BPCAT), we showed how models can be configured and how simulation can be performed based on specific objectives. The results of this case study suggest that the storage volume required for storing blockchain data can be controlled and even reduced through the use of off-chain storage, depending on the difference rates in the data. It should be noted that in manufacturing processes, a machine can often produce similar traceability data, suggesting a low deviation rate and a real opportunity for reducing the data storage volume. In addition, we showed that energy consumption, which is an issue that manufacturers often use as an argument against blockchain solutions, could also be controlled by judicious distribution among blockchain participants. In fact, each participant actually profits from the immutability and transparency provided by blockchain. Consequently, instead of charging the manufacturing plant for all the energy costs as well as the environmental impact, it is possible to distribute them across the blockchain participants, for example, as a service offered to suppliers and customers who subscribe.

The simulator generated from the modeling can be used for evaluating the plant over periods ranging from a few days to years of production, further allowing actual production data to be used. In our future research, we plan to extend the framework so that it can be used as an operational management or deployment/development planning tool in Factory 4.0.

4.9. References

Alkhader, W., Alkaabi, N., Salah, K., Jayaraman, R., Arshad, J., Omar, M. (2020). Blockchain-based traceability and management for additive manufacturing. *IEEE Access*, 8, 188363–188377.

Bergman, S. (2020). How can I calculate CO2eq emissions for my Azure VM? [Online]. Available at: https://devblogs.microsoft.com/sustainable-software/how-can-i-calculate-co2eq-emissions-for-my-azure-vm.

Bodkhe, U., Tanwar, S., Parekh, K., Khanpara, P., Tyagi, S., Kumar, N., Alazab, M. (2020). Blockchain for industry 4.0: A comprehensive review. *IEEE Access*, 8, 79764–79800.

Chen, S., Cai, X., Wang, X., Liu, A., Lu, Q., Xu, X., Tao, F. (2022). Blockchain applications in PLM towards smart manufacturing. *Int. J. Adv. Manuf. Technol.*, 118, 2669–2683.

De Giovanni, P. (2020). Blockchain and smart contracts in supply chain management: A game theoretic model. *International Journal of Production Economics*, 228, 107855 [Online]. Available at: https://www.sciencedirect.com/ science/article/pii/S0925527320302188.

Hader, M., Tchoffa, D., Mhamedi, A.E., Ghodous, P., Dolgui, A., Abouabdellah, A. (2022). Applying integrated blockchain and big data technologies to improve supply chain traceability and information sharing in the textile sector. *Journal of Industrial Information Integration*, 28, 100345 [Online]. Available at: https://www.sciencedirect.com/ science/article/pii/S2452414X22000176.

Hasan, H.R., Salah, K., Jayaraman, R., Ahmad, R.W., Yaqoob, I., Omar, M. (2020). Blockchain-based solution for the traceability of spare parts in manufacturing. *IEEE Access*, 8, 100308–100322.

Huo, R., Zeng, S., Wang, Z., Shang, J., Chen, W., Huang, T., Wang, S., Yu, F.R., Liu, Y. (2022). A comprehensive survey on blockchain in industrial internet of things: Motivations, research progresses, and future challenges. *IEEE Communications Surveys & Tutorials*, 24(1), 88–122.

Ko, T., Lee, J., Ryu, D. (2018). Blockchain technology and manufacturing industry: Real-time transparency and cost savings. *Sustainability*, 10(11) [Online]. Available at: https://www.mdpi.com/2071-1050/10/11/4274.

Lee, J., Azamfar, M., Singh, J. (2019). A blockchain enabled Cyber-Physical System architecture for Industry 4.0 manufacturing systems. *Manufacturing Letters*, 20, 34–39 [Online]. Available at: https://www.sciencedirect.com/science/article/pii/S2213846319300264.

Leng, J., Ruan, G., Jiang, P., Xu, K., Liu, Q., Zhou, X., Liu, C. (2020). Blockchain-empowered sustainable manufacturing and product lifecycle management in Industry 4.0: A survey. *Renewable and Sustainable Energy Reviews*, 132, 110112 [Online]. Available at: https://www.sciencedirect.com/science/article/ pii/S1364032120304032.

Leng, J., Ye, S., Zhou, M., Zhao, J.L., Liu, Q., Guo, W., Cao, W., Fu, L. (2021). Blockchain-secured smart manufacturing in Industry 4.0: A survey. *IEEE Transactions on Systems, Man, and Cybernetics: Systems*, 51(1), 237–252.

Li, Z., Barenji, A.V., Huang, G.Q. (2018). Toward a blockchain cloud manufacturing system as a peer to peer distributed network platform. *Robotics and Computer-Integrated Manufacturing*, 54, 133–144 [Online]. Available at: https://www.sciencedirect.com/science/article/pii/S073658451830022X.

Mullet, V., Sondi, P., Ramat, E. (2021). A review of cybersecurity guidelines for manufacturing factories in Industry 4.0. *IEEE Access*. 9, 23235–23263.

Mullet, V., Sondi, P., Ramat, E. (2022). Enhancing trust in Industry 4.0 traceability data using confidentiality-preserving digital ledger. In *Proceedings of the 4th Conference on Blockchain Research & Applications for Innovative Networks and Services (BRAINS)*. IEEE, Paris.

Ramat, E. (2022). Artis*: Devs multimodeling and simulation framework [Online]. Available at: https://gitlab.com/artis-star/artis-star.

Tönnissen, S., Teuteberg, F. (2020). Analysing the impact of blockchain-technology for operations and supply chain management: An explanatory model drawn from multiple case studies. *International Journal of Information Management*, 52, 101953.

Wang, Y. and Kogan, A. (2018). Designing confidentiality-preserving blockchain-based transaction processing systems. *International Journal of Accounting Information Systems*, 30, 1–18 [Online]. Available at: https://www.sciencedirect. com/science/article/pii/S1467089518300794.

Zeigler, B. (1976). *Theory of Modeling and Simulation*. Elsevier, Amsterdam.

5

Digitalization-Centered Perspectives for Factory 4.0

The research presented throughout this book addresses the integration of a blockchain solution into the factory 4.0 traceability system with the underlying aim of fostering trust between partners without jeopardizing the privacy of traceability data.

5.1. Toward a safer and more secure digitalization of factories

We first identified what the state of the art was for Factory 4.0 from a cybersecurity perspective in order to provide a better general overview of existing technologies and systems. After introducing the concept of Industry 4.0, we demonstrated the complexity presented by the mechanisms underlying cybersecurity to ensure system security due to the large number of new technologies involved. We then proposed a characterization of cybersecurity in terms of both its technical and managerial characteristics to show that every level in an organization has a role to play. Furthermore, we emphasized that cybersecurity could also be leveraged for value creation.

5.1.1. *Contributions of the book to factory 4.0 cybersecurity*

Since cybersecurity must be taken into consideration based on physical zones, we proposed a division of a factory into several perimeters, which we characterized according to their interactions, equipment and networks. We then reported a list of vulnerabilities, threats and risks in cybersecurity met in Industry 4.0 factories for each perimeter, while accounting for the business impacts at the organizational level. It should be noted that in addition to the paradigm underlying the expert system usually implemented in cybersecurity solutions, new innovative technologies such as machine learning, *honeypots* and digital twins, are used in recent solutions.

Cybersecurity and Traceability in Factory 4.0,
coordinated by Patrick SONDI and Valentin MULLET.

We completed this review of scientific solutions with a summary of good practices promoted by official bodies which can be used to lay the first building block for establishing a cybersecurity strategy. This significant contribution (Mullet et al. 2021) has been referenced in more than a hundred scientific publications on the industry of the future in three years.

Our analysis of all this referenced work as well as the standards and recommendations of the main organizations is that cybersecurity is still implemented today as an activity intended to detect vulnerabilities in existing systems and to address them. This gives the impression that a big race is taking place behind the attacks that, in most cases, appear to be inevitable and as such we must strive to delay them or repair their consequences. More than ever, an interesting standpoint involves the rethinking of the design and implementation of systems by integrating security, and therefore cybersecurity. These should be an essential feature of their normal operation and the choice of techniques and tools that must be integrated into the system with the imperative constraint of preserving this characteristic.

5.1.2. *Contributions of the book to blockchain for traceability*

A first attempt at this strategy takes the form of our blockchain-based traceability solution for manufacturing plants, which aims to offer an effective way to ensure transparency to all partners involved in a supply chain while protecting the privacy of their respective critical data (Mullet et al. 2023).

The main idea is to include confidential data that were encrypted through specific mechanisms along with its hash, computed before encryption, into the transactions stored in the blockchain. Therefore, in the event of a dispute, investigators can request the data owner access the confidential information and compare them with the related data validated by all the parties participating in the blockchain in order to establish their authenticity during the analysis procedure for finding product defects.

It should be noted that homomorphic cryptography is a promising avenue for resolving the potential deadlock that the refusal of the originator of the data to decrypt them may cause. The mechanisms previously listed actually make it possible to directly verify the encrypted data without the need to decrypt them. In any case, encryption will still be required before inserting the data into the blockchain for the purposes of protecting confidentiality. In order to precisely mitigate any delay related to encryption/decryption operations and avoid prohibitive storage use for traceability files, only the file hash must be included in the transaction stored in the blockchain. On the other hand, the data can be stored by other devices described in detail in this book. The implementation of the concepts and functionalities proposed was illustrated through the use of the MultiChain *open source* blockchain platform. We showed that the storage of lineage files can be optimized to reduce the data volume and the total storage volume needed by each node by using multichain native *off-chain* storage and stream functions.

5.1.3. *Perspectives*

An interesting future avenue for our work would be to explore the possibility of integrating approaches such as homomorphic cryptography (Gong et al. 2024) and zero-knowledge proof (Sun et al. 2021) in the design of blockchain-based traceability systems and natively supporting confidentiality.

5.2. Toward more dynamic management using digital twins

We took into consideration that a large number of manufacturers hesitate in deploying blockchain mainly due to the circulation of information about many characteristics that have a deterring effect. This is particularly the case in terms of energy consumption and storage. Therefore, bearing in mind the feedback primarily involving this information, we also propose a novel approach for the a priori evaluation of this type of project.

5.2.1. *Contributions of the book to the evaluation of blockchain projects*

Based on the DEVS paradigm for the specification of discrete event systems, our main proposal consists of a collaborative modeling framework of a manufacturing plant, in which each component can be modeled and configured by an expert as finely as necessary according to the evaluation goals. The framework also provides an overview of the plant where managers can configure (production planning, storage management, etc.) and observe patterns (storage volume, energy consumption, etc.).

Applying this approach to our proposal for a blockchain-based traceability system in an Industry 4.0 manufacturing plant (BPCAT), we showed how models can be configured and how simulation can be performed based on specific objectives.

The results of this case study suggest that the volume of storage needed to store blockchain data can be controlled and even reduced when using off-chain storage, depending on the difference rate in the data. It should be noted that in manufacturing processes, a machine can often produce similar traceability data, suggesting a low deviation rate and therefore a real possibility for reducing the data storage volume. In addition, we showed that energy consumption, which is an issue often used by manufacturers to argue against blockchain-based solutions, can also be brought under control, on the one hand by an appropriate choice of consensus algorithm and, on the other hand, by an appropriate distribution among the blockchain participants. Furthermore, each participant may take advantage of the immutability and transparency provided by the blockchain. Consequently, instead of charging the manufacturing plant for the full energy and environmental impact-related costs, these can be distributed among the blockchain participants, as a service offered to suppliers and customers who subscribe thereto.

5.2.2. *Prospects*

The simulator generated from the proposed modeling can be used to evaluate the plant over different periods of time such as several days, weeks or even years of production. Until now, the simulator has been mainly used for evaluating the impact of implementing a blockchain in a factory in which the BPCAT approach was adopted, in particular using real production data that was previously anonymized and injected into the simulator in the form of production orders.

Industry 4.0 introduces the notion of digital twin that represents a digital replica of the entire physical factory infrastructure. The companies that deploy it primarily approach it as a means to collect massive amounts of real-time data to satisfy immediate operational requirements or to recreate production scenarios in order to understand what happened. However, another major benefit of the digital twin is that it can be utilized to evaluate any possible alteration envisaged to the plant and to understand all the impacts in order to only implement them in the real production system (PS) if these will bring about any advantage and can be done under controlled conditions.

We have undertaken research work (Batchoudi and Ramat 2024) that aims at introducing into the achieved modeling of plant simulation different optimization models used in factory management. In this optimization and simulation coupling approach, the data provided by the numerical twin could be used to determine several scenarios of different possible developments, to which the appropriate optimizations would be applied while at the same time the simulation would enable the associated impacts to be evaluated.

An interesting project would consist of extending the proposed models so that they can be used as a management and operational planning tool for the development of technological solutions in a Factory 4.0 before their implementation in an actual factory.

5.3. Toward digitalization while ensuring factory resilience

As mentioned earlier, the development of high-speed Internet has led to the emergence of cloud computing, which makes possible pooling storage and processing resources for increasingly large volumes of data. Already highly automated, industrial production with Industry 4.0 is no exception to this trend and is entering this new era of intense digitalization and connectivity. Therefore, in addition to the initiatives from manufacturers and standardization organizations, many academic works have focused on proposing frameworks to drive the transformation of PSs and the associated information system into a cyber-physical production system (CPPS) (Wu et al. 2021) toward Industry 4.0.

5.3.1. *Assessment of the state of factory resilience to cope with digitalization*

The last few years have seen communication technologies converge toward an adapted solution to any kind of use. This has been possible by offering increasingly higher speeds and wider coverage, and large Internet companies have invested massively in cloud infrastructures driven by the ambition of offering shared hosting to all applications and all data. In addition, the advent of artificial intelligence-based techniques has finally allowed for efficiently processing massive amounts of data for all kinds of optimization purposes. Thus, all the signs have indicated that this industrial revolution could calmly initiate and foster the transformation of factories.

Consequently, the question of plant resilience and its ability to maintain its operation in the event of failure of all or part of these transformation elements is often addressed as a transitory risk, but not from the perspective of a well-framed and sustainable return of the PS to a state prior to all or part of these transformations (Attajer et al. 2022).

Nonetheless, it is clearly visible that this transformation, which involves the integration of components 4.0 into the initial PS (such as the Internet of Things (IoT) for data collection, remote control and predictive maintenance, traceability-based blockchain as illustrated in this book, cloud for data storage and Big Data services needed for service optimization, etc.), exposes the PS to increased vulnerabilities and risks.

These new components 4.0 have introduced in factories risks associated with new technologies. They are responsible for an increased dependency on the Internet, which in the event of interruptions inexorably leads to losing access to the cloud, and as such may cause production downtimes. They also have caused massive exposure to cybersecurity risks through their connection to the Internet and the introduction into production lines of connected objects requiring a low level of administration and security, as we pointed out in the first chapter of this book.

Against the backdrop of the current context of the climate, energy and geopolitical crises, the consequences of this digitalization should furthermore include increased energy consumption and internal carbon footprint of the plant, the dependency on cloud services operated mainly by non-national players, as well as the increased pressure on raw materials and electronic components that may cause some components 4.0 to become unrepairable or unavailable.

Most of the operational proposals approach resilience from the specific standpoint of cybersecurity (Espinoza-Zelaya and Moon 2021), while those formulated by institutions instead focus on data sovereignty in the cloud (Gaia-X 2021).

Yet, very few studies address resilience as a broader topic that could lead to a thorough transformation of the cyber-physical PS, even going as far as to reverse previous measures and that would involve the removal of certain components 4.0. Nonetheless, this seems to be the ultimate solution, especially in response to energy and raw material shortages. It was observed, for example, when massive increases occurred in energy prices as a result of the context of the war in Ukraine in 2022, that several companies had to start business continuity plans in order to try to maintain as much as possible a certain activity level in factories, when other factories were simply going out of business for an indefinite period. Unfortunately, in the absence of a methodological framework and real upstream analyses on resilience, at least at the technological level, these continuity plans are put into practice in haste without any visibility on the real impact on all the plant systems, nor on the sustainability of the measures implemented, and even less on the new situation faced by the plant concerning its legal, normative, contractual and quality-wise obligations.

A first study that really tries to address resilience at the most global level possible is proposed in Goepp et al. (2022). An attempt is made by the authors to match resilience characteristics, phases and strategies to the various components 4.0. Nevertheless, weaknesses and risks are still addressed as being exceptional and not as sustainable, as we pointed out when we approached the energy and geopolitical risks or those involving raw materials; this will inevitably lead back to the current transitional solutions in terms of resilience.

5.3.2. *Prospects*

We have started research work for developing an approach that integrates both methodological contributions as well as technical considerations and solutions that from the outset include plant resilience in the transformation process of its PS into a cyber-physical PS. The idea is to define at each transformation stage the aspects that should be taken into account so that the integration of any component 4.0 comprises a definition of the conditions under which the component can be integrated, maintained or excluded from the plant operation according to the risks involved. In order to implement these operations within a controlled framework, this definition must be accompanied by a description of the impacts associated with the decision taken for each particular component 4.0 and the consequences of these impacts on the situation of the plant concerning its legal, normative, contractual and quality constraints. As a result, a company becomes capable of implementing and developing in a relatively simple manner a business continuity plan within a controlled framework, independently of whether the vulnerabilities and risks involved are temporary or long-term ones.

To demonstrate the concept of the proposed methodology and to allow manufacturers to visualize the potential of this approach for their factory, we defined a case study (Hostos et al. 2024) on the Industry 4.0 component associated with the IoT. Given the role that industrial IoT (IIoT) plays in the traceability of the Factory 4.0, the connection between this future research and that presented in this book will be quite obvious.

5.4. References

Attajer, A., Chaabane, S., Darmoul, S., Sallez, Y., Riane, F. (2022). Evaluation of operational resilience in cyber-physical production systems Literature review. *IFAC-PapersOnLine*, 55(10), 2264–2269 [Online]. Available at: https//www.sciencedirect.com/science/article/pii/S2405896322020547.

Batchoudi, G. and Ramat, E. (2024). Transformation d'un JSON-DSL vers un couplage Parallel-DEVS-UML State machine pour les systèmes de production. In *Journées Francophones de la Modélisation et de la Simulation (JFMS)*, Cargèse.

Espinoza-Zelaya, C., and Moon, Y. (2021). Resilient cyber-manufacturing systems under cyber attacks. In *Proceedings of the ASME 2021 International Mechanical Engineering Congress and Exposition. Volume 2B: Advanced Manufacturing*, 1–5 November, V02BT02A011.

Gaia-X (2021). Gaia-X architecture document [Online]. Available at: https//gaia-x.eu/.

Goepp, V., Berrah, L., Drira, E., Chaabane, S. (2022). Towards a framework to position resilience and Industry 4.0. *IFAC-PapersOnLine*, 55(10), 2258–2263.

Gong, Y., Chang, X., Mišić, J., Mišić, V.B., Wang, J., Zhu, H. (2024). Practical solutions in fully homomorphic encryption A survey analyzing existing acceleration methods. *Cybersecurity*, 7(1), 5. doi: 10.1186/s42400-023-00187-4.

Hostos, H., Goepp, V., Sondi, P. (2025). A framework for resilient integration of Industry 4.0 components into production systems. In *Innovative Intelligent Industrial Production and Logistics*, Dassisti, M., Madani, K., Panetto, H. (eds). Springer, Cham.

Mullet, V., Sondi, P., Ramat, E. (2021). A review of cybersecurity guidelines for manufacturing factories in Industry 4.0. *IEEE Access*, 9, 23235–23263.

Mullet, V., Sondi, P., Ramat, E. (2023). A blockchain-based confidentiality-preserving approach to traceability in Industry 4.0. *Int J Adv Manuf Technol*, 124, 1297–1320.

Sun, X., Yu, F.R., Zhang, P., Sun, Z., Xie, W., Peng, X. (2021). A survey on zero-knowledge proof in blockchain. *IEEE Network*, 35(4), 198–205.

Wu, X., Goepp, V., Siadat, A., Vernadat, F. (2021). A method for supporting the transformation of an existing production system with its integrated Enterprise Information Systems (EIS) into a Cyber Physical Production System (CPPS). *Computers in Industry*, 131, 103483.

List of Authors

Valentin MULLET
LISIC
ULCO
Calais
France

Patrick SONDI
CERI Systèmes Numériques
IMT Nord Europe
Villeneuve-d'Ascq
France

Index

L, M

N

O, P

Q, R

S

T

U, V

Printed and bound by CPI Group (UK) Ltd, Croydon, CR0 4YY
13/09/2026
14961074-0001